Radio Frequency Transmission Systems

Design and Operation

Other McGraw-Hill Reference Books of Interest

Handbooks

Avalone and Baumeister • STANDARD HANDBOOK FOR MECHANICAL ENGINEERS
Beeman • INDUSTRIAL POWER SYSTEMS HANDBOOK
Benson • AUDIO ENGINEERING HANDBOOK
Benson • TELEVISION ENGINEERING HANDBOOK
Benson and Whitaker • TV AND AUDIO HANDBOOK
Coombs • BASIC ELECTRONIC INSTRUMENT HANDBOOK
Coombs • PRINTED CIRCUITS HANDBOOK
Croft and Summers • AMERICAN ELECTRICIANS' HANDBOOK
Di Giacomo • VLSI HANDBOOK
Fink and Beaty • STANDARD HANDBOOK FOR ELECTRICAL ENGINEERS
Fink and Christiansen • ELECTRONICS ENGINEERS' HANDBOOK
Hicks • STANDARD HANDBOOK OF ENGINEERING CALCULATIONS
Inglis • ELECTRONIC COMMUNICATIONS HANDBOOK
Johnson and Jasik • ANTENNA ENGINEERING HANDBOOK
Juran • QUALITY CONTROL HANDBOOK
Kaufman and Seidman • HANDBOOK FOR ELECTRONICS ENGINEERING TECHNICIANS
Kaufman and Seidman • HANDBOOK OF ELECTRONICS CALCULATIONS
Kurtz • HANDBOOK OF ENGINEERING ECONOMICS
Mee and Daniel • MAGNETIC RECORDING HANDBOOK
Sherman • CD-ROM HANDBOOK
Stout • HANDBOOK OF MICROPROCESSOR DESIGN AND APPLICATIONS
Stout and Kaufman • HANDBOOK OF MICROCIRCUIT DESIGN AND APPLICATION
Stout and Kaufman • HANDBOOK OF OPERATIONAL AMPLIFIER DESIGN
Tuma • ENGINEERING MATHEMATICS HANDBOOK
Williams • DESIGNER'S HANDBOOK OF INTEGRATED CIRCUITS
Williams and Taylor • ELECTRONIC FILTER DESIGN HANDBOOK

Consumer Electronics

Luther • Digital VIDEO IN THE PC ENVIRONMENT
Mee and Daniel • MAGNETIC RECORDING, VOLUMES I-III
Philips International • COMPACT DISC—INTERACTIVE

Dictionaries

DICTIONARY OF COMPUTERS
DICTIONARY OF ELECTRICAL AND ELECTRONIC ENGINEERING
DICTIONARY OF ENGINEERING
DICTIONARY OF SCIENTIFIC AND TECHNICAL TERMS
Markus • ELECTRONICS DICTIONARY

Radio Frequency Transmission Systems
Design and Operation

Jerry C. Whitaker

Intertext Publications
McGraw-Hill Book Company

New York St. Louis San Francisco Auckland Bogotá
Hamburg London Madrid Mexico Milan Montreal
New Delhi Panama Paris São Paolo
Singapore Sydney Tokyo Toronto

Library of Congress Catalog Card Number 90-81425

Whitaker, Jerry C.
   Radio frequency transmission systems : design and operation /
Jerry C. Whitaker.
      434 p.
   Includes bibliographical references and index
   ISBN 0-07-069620-9
   1. Radio--Transmitters and transmission.   I. Title
TK6561.W49 1990
621.384' 15--dc20   90-6320 CIP

1 2 3 4 5 6 7 8 9 0 XXX/XXX 9 5 4 3 2 1 0

ISBN 0-07-069620-9

The sponsoring editor for this book was Daniel A. Gonneau.
Cover Photo: Lanny Whitaker, The Weather Channel, Atlanta, GA
Art Direction: Kristi Sherman
Provided courtesy of Intertec Publishing Corp.

Printed and bound by
Intertext Publications/Multiscience Press, Inc.
One Lincoln Plaza
New York, NY 10023

This book is dedicated to my dad,
Clyde E. Whitaker

Thank you for the guidance, patience and understanding.

Contributors and Reviewers

The author wishes to express appreciation to the following individuals who provided input and comments on this publication.

Dr. John H. Battison, P.E., Consulting Engineer, Loudonville, OH.

Carl A. Bentz, Intertec Publishing Corp., Overland Park, KS.

Earl Blankenship, Varian Associates, Palo Alto, CA.

Jerry Collins, Harris Corporation, Quincy, IL.

Bradley Dick, Intertec Publishing Corp., Overland Park, KS.

Donald Markley, P.E., Markley and Associates Consultants, Peoria, IL.

Earl McCune, Varian Associates, Palo Alto, CA.

Robert Perelman, Andrew Corporation, Orland Park, IL.

Merrald B. Shrader, Eimac Division, Varian Associates, San Carlos, CA.

Hilmer Swanson, Harris Corporation, Quincy, IL.

Contents

Preface

Radio frequency (RF) transmission is one of the oldest forms of electronics. From the days of Hertz and Marconi, RF transmission pioneered the art and science of electrical engineering. It served as the basis for a myriad of related applications, not the least of which includes audio amplification and processing, video pickup and reproduction, and radar. RF technology has reshaped our national defense efforts, radically changed the way we communicate, provided new products and services, and brought nations together to celebrate good times and mourn bad times. RF is an invisible technology. It is a discipline that is seldom taught in colleges and universities. Yet, radio frequency transmission equipment has reshaped the way we live.

This book is intended to serve the information needs of persons who specify, install, and maintain RF equipment. The wide variety of hardware currently in use requires that personnel involved in RF work be familiar with a multitude of concepts and applications. This publication examines a wide range of technologies and power devices focusing on devices and systems that produce in excess of 1 kW.

Extensive theoretical dissertations and mathematical explanations have been included only to the extent that they are essential for an understanding of the basic concepts. Excellent reference books are available from this publisher that examine individual RF devices and their underlying design criteria. This book puts the individual elements together and shows how they interrelate.

The areas covered by *Radio Frequency Transmission Systems* range from broadcasting to electronic countermeasures. The basic concepts and circuit types of all major RF applications are covered. Generous use of illustrations makes difficult or complex concepts easier to comprehend and practical examples are provided wherever possible. Special emphasis is given to radio and television hardware because these applications provide examples that may be readily translated to other uses. While in-depth theory and math is not included in this publication, a thorough understanding of basic electrical concept is assumed by the author.

The *Radio Frequency Transmission Systems* handbook is divided into the following major subject areas:

- **Applying RF Technology** (Chapters 1 through 3). Common uses of radio frequency energy are examined, and examples given. Chapter 1 provides an overview of the frequency bands, modulation methods, and amplifier operating classes. Chapter 2 examines RF application in broadcasting. Chapter 3 examines RF applications in satellite communications, radar, and distance-direction finding.
- **Solid State Power Devices** (Chapters 4 and 5). The operating parameters of semiconductor-based power devices are discussed, and examples of typical circuits are given. Chapter 4 outlines the basic principles of bipolar and FET semiconductors, including potential failure modes. Chapter 5 presents practical examples of RF amplifiers based on solid state devices, including a digitally modulated AM transmitter.
- **Power Vacuum Tube Devices** (Chapters 6 and 7). The basic principles and applications of gridded vacuum tubes are outlined and example circuits provided. Chapter 6 examines the theory behind vacuum tube operation, including neutralization and potential failure modes. Chapter 7 provides numerous examples of RF systems that use vacuum tubes as the primary energy-generating element.
- **Microwave Power Tubes** (Chapters 8 and 9). The operating principles of classic microwave devices and new high-efficiency tubes are given. Chapter 8 reviews the basic concepts of klystrons, traveling-wave tubes and other microwave power devices. Chapter 9 examines high-efficiency tubes that have recently become available, including the Klystrode tube and multistage depressed collector klystron. Both devices offer significant efficiency benefits when operated in linear service.
- **RF Components and Transmission Line** (Chapters 10 and 11). The operation of hardware used to combine and conduct RF power is explained. Chapter 10 examines coaxial transmission line and waveguide. Chapter 11 explains the operation of RF combining systems, including hot-switches and circulators.
- **Antenna Systems** (Chapter 12). An overview of antenna types is given and basic operating parameters described. Examples are provided of antennas used in radio and TV broadcasting.

RF power technology is a complicated, but exciting science. It is a science that advances each year. The frontiers of higher power and higher frequency continue to fall as new applications drive new developments by manufacturers. Radio frequency technology is not an aging science. It is a discipline that is, in fact, just reaching its stride.

Using RF Power Technology

1.1 Introduction

Radio frequency (RF) power amplifiers are used in countless applications at tens of thousands of facilities around the world. This wide variety of applications, however, stems from a few basic concepts of conveying energy and information by means of a radio frequency signal. Further, the devices used to produce RF energy have many similarities, regardless of the final application. Although radio and television broadcasting represent the most obvious use of high-power RF generators, numerous other common applications exist, including:

- Induction heating and process control systems
- Radio communications (2-way mobile radio base stations and cellular base stations)
- Amateur radio
- Radar (ground, air, and shipboard)
- Satellite communications
- Atomic science research
- Medical research, diagnosis, and treatment

1.1.1 Electromagnetic-Radiation Spectrum

The usable spectrum of electromagnetic-radiation frequencies extends over a range starting below 100 Hz for power distribution and increasing up to 10^{20} Hz for the shortest X-rays (see Figure 1-1). The lower frequencies are used primarily for terrestrial broadcasting and

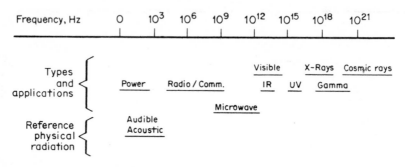

Figure 1-1 The electromagnetic spectrum. (*Source:* Fink, 1982.)

communications. The higher frequencies include visible and near-visible infrared and ultraviolet light, and X-rays. The frequencies of interest to RF engineers range from 30 kHz to 30 GHz.

Low Frequency (LF): 30 kHz to 300 kHz. The LF band is used for around-the-clock communications services over long distances, and where adequate power is available to overcome high levels of atmospheric noise. Applications include:

• Radionavigation
• Fixed/maritime communications
• Aeronautical radionavigation
• Low frequency broadcasting (Europe)
• Underwater submarine communications (10-30kHz)

Medium Frequency (MF): 300 kHz to 3 MHz. The low frequency portion of this band is used for around-the-clock communication services over moderately long distances. The upper portion of the MF band is used principally for moderate distance voice communications. Applications include:

• AM radio broadcasting (535.5-1605.5 kHz)
• Radionavigation
• Fixed/maritime communications
• Aeronautical radionavigation
• Fixed and mobile commercial communications
• Amateur radio
• Standard time and frequency services

High Frequency (HF): 3 MHz to 30 MHz. This band provides reliable medium-range coverage during daylight and, when the transmission path is in total darkness, worldwide long-distance service. (The reliability and signal quality of long-distance service depends to a large degree upon ionospheric conditions and related long-term variations in sun-spot activity affecting sky-wave propagation.) Applications include:

- Shortwave broadcasting
- Fixed and mobile service
- Telemetry
- Amateur radio
- Fixed/maritime mobile
- Standard time and frequency services
- Radio astronomy
- Aeronautical fixed and mobile

Very High Frequency (VHF): 30 MHz to 300 MHz. The VHF band is characterized by reliable transmission over medium distances. At the higher portion of the VHF band, communications is limited by the horizon. Applications include:

- FM radio broadcasting (88-108 MHz)
- Low band VHF television broadcasting (54-72 MHz and 76-88 MHz)
- High band VHF television broadcasting (174-216 MHz)
- Commercial fixed and mobile radio
- Aeronautical radionavigation
- Space research
- Fixed/maritime mobile
- Amateur radio
- Radiolocation

Ultra High Frequency (UHF): 300 MHz to 3 GHz. Transmissions in this band are typically line-of-sight. Short wavelengths at the upper end of the band permit the use of highly directional parabolic or multielement antennas. Applications include:

- UHF terrestrial television (470-806 MHz)
- Fixed and mobile communications
- Telemetry
- Meteorological aids
- Space operations

- Radio astronomy
- Radionavigation
- Satellite communications
- Point-to-point microwave relay

Super High Frequency (SHF): 3 GHz to 30 GHz. Communications in this band are strictly line-of-sight. Very short wavelengths permit the use of parabolic transmit and receive antennas of exceptional gain. Applications include:

- Satellite communications
- Point-to-point wideband relay
- Radar
- Specialized wideband communications
- Developmental research
- Military support systems
- Radiolocation
- Radionavigation
- Space research

1.2 Modulation Systems

The primary purpose of most communications systems is to transfer information from one location to another. The message signals used in communications and control systems usually must be limited in frequency to provide for efficient transfer. This frequency may range from a few hertz for control systems to a few megahertz for video signals. To facilitate efficient and controlled distribution of these signals, an *encoder* is generally required between the source and the transmission channel. The encoder acts to *modulate* the signal, producing at its output the *modulated waveform*. Modulation is a process whereby the characteristics of a wave (the *carrier*) are varied in accordance with a message signal, the modulating waveform. Frequency translation is usually a byproduct of this process. Modulation can be continuous — where the modulated wave is always present, or pulsed — where no signal is present between pulses.

There are a number of reasons for producing modulated waves, including:

- Frequency Translation. The modulation process provides a vehicle to perform the necessary frequency translation required for distribution of information. An input signal may be translated to its assigned frequency band for transmission or radiation.

- Signal processing. It is often easier to amplify or process a signal in one frequency range as opposed to another.
- Antenna Efficiency. Generally speaking, for an antenna to be efficient, it must be large compared to the signal wavelength. Frequency translation provided by modulation allows antenna gain and beamwidth to become part of the system design considerations. Use of higher frequencies permits antenna structures of reasonable size and cost.
- Bandwidth Modification. The modulation process permits the bandwidth of the input signal to be increased or decreased as required by the application. Bandwidth reduction can permit more efficient use of the spectrum, at the cost of signal fidelity. Increased bandwidth, on the other hand, permits increased immunity to transmission channel disturbances.
- Signal Multiplexing. In a given transmission system it may be necessary or desirable to combine several different signals into one baseband waveform for distribution. Modulation provides the vehicle for such *multiplexing*. Various modulation schemes allow separate signals to be combined at the transmission end, and separated (*demultiplexed*) at the receiving end. Multiplexing may be accomplished by using *frequency-domain multiplexing* (FDM) or *time-domain multiplexing* (TDM).

Modulation of a signal does not come without undesirable characteristics. Bandwidth restriction or the addition of noise and/or other disturbances are the two primary problems faced by the transmission system designer.

1.3 Amplitude Modulation

In the simplest form of amplitude modulation, an analog carrier is controlled by an analog modulating signal. The desired result is an RF waveform whose amplitude is varied by the magnitude of the audio signal, and at a rate equal to the frequency of the audio signal. The resulting waveform consists of a carrier wave plus two additional signals: an upper-sideband signal, which is equal in frequency to the carrier *plus* the frequency of the modulating signal; and a lower-sideband signal, which is equal in frequency to the carrier *minus* the frequency of the modulating signal. This type of modulation system is referred to as *double-sideband amplitude modulation* (DSAM).

Amplitude modulation (AM) is essentially a multiplication process in which the time functions that describe the modulating signal and the carrier are multiplied together to produce a modulated wave containing *intelligence* (information or data of some kind). The frequency components of the modulating signal are translated in this process to occupy a different position in the spectrum. The bandwidth of an AM transmission is determined by the modulating frequency. The bandwidth required for full fidelity reproduction in a receiver is equal to twice the applied modulating frequency.

The magnitude of the upper sideband and lower sideband will not normally exceed 50 percent of the carrier amplitude during modulation. This results in an upper-sideband power of one-fourth the carrier power. The same power exists in the lower sideband. As a result, up to one half of the actual carrier power appears additionally in the sum of the sidebands of the modulated signal. A representation of the AM carrier and its sidebands is shown in Figure 1-2. The actual occupied bandwidth, assuming pure sinusoidal modulating signals and no distortion during the modulation process, is equal to twice the frequency of the modulating signal. Full (100 percent) modulation occurs when the peak value of the modulated envelope reaches twice the value of the unmodulated carrier, and the minimum value of the envelope is zero. The envelope of a modulated AM signal in the time domain is shown in Figure 1-3.

When modulation exceeds 100 percent on the negative swing of the carrier, spurious signals are emitted. It is possible to modulate an AM carrier asymmetrically; that is, to restrict modulation in the negative direction to 100 percent, but to allow modulation in the positive direction to exceed 100 percent without a significant loss of fidelity.

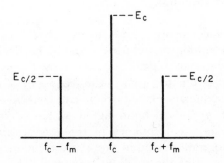

Figure 1-2 Frequency-domain representation of an amplitude modulated signal at 100 percent modulation. E_c = carrier power, F_c = frequency of the carrier, and F_m = frequency of the modulating signal. (*Source:* K. Blair Benson, *Audio Engineering Handbook,* McGraw-Hill, New York, 1988.)

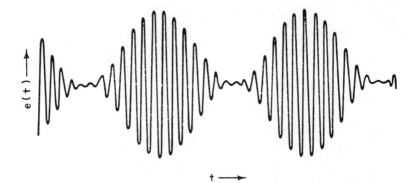

Figure 1-3 Time-domain representation of an amplitude-modulated signal. Modulation at 100 percent is defined as the point at which the peak of the waveform reaches twice the carrier level, and the minimum point of the waveform is zero. (*Source:* K. Blair Benson, *Audio Engineering Handbook*, McGraw-Hill, New York, 1988.)

In fact, many modulating signals normally exhibit asymmetry, most notably human speech waveforms.

1.3.1 Vestigial-Sideband Amplitude Modulation

Because the intelligence (modulating signal) of a conventional AM transmission is identical in the upper *and* lower sidebands, it is possible to eliminate one sideband and still convey the required information. This scheme is implemented in *vestigial-sideband AM* (VSBAM). Complete elimination of one sideband (for example, the lower sideband) requires an ideal high-pass filter with an infinitely sharp cutoff. Such a filter is difficult to implement in any practical design. VSBAM is a compromise technique wherein one sideband (typically the lower sideband) is attenuated significantly. The result is a savings in occupied bandwidth and transmitter power.

VSBAM is used for television broadcast transmission. A typical bandwidth trace for a VSBAM TV transmitter is shown in Figure 1-4.

1.3.2 Single-Sideband Amplitude Modulation

If the necessary filtering is provided in the transmitter, *single-sideband AM* (SSBAM) is practical, and results in significant savings in occupied bandwidth and transmitter power. SSBAM is used extensively for fixed and mobile communications.

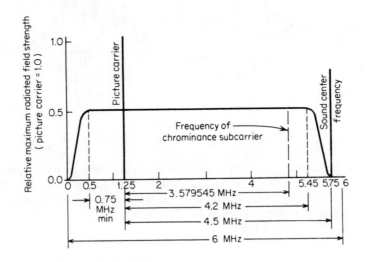

Figure 1-4 Idealized amplitude characteristics of the FCC standard waveform for monochrome and color television transmission. (From FCC Rules, Section 73.699.)

1.3.3   Suppressed-Carrier AM

The carrier in an AM signal does not convey any intelligence. All of the modulating information is in the sidebands. It is possible, therefore, to suppress the carrier upon transmission, radiating only the sidebands of the AM signal. The result is much greater efficiency at the transmitter (that is, a reduction in the required transmitter power). Suppression of the carrier may be done with DSAM and SSBAM signals. *Single-sideband suppressed carrier AM* (SSB-SC) is the most spectrum and energy efficient mode of AM transmission. Figure 1-5 shows waveforms for suppressed carrier transmissions.

The drawback to suppressed-carrier systems is the requirement for a more complicated receiver. The carrier must be regenerated at the receiver to permit demodulation of the signal. Also, in the case of SSBAM transmitters, it is usually necessary to generate the SSB signal in a low-power stage and then amplify the signal with a linear-power amplifier to drive the antenna. Linear amplifiers generally exhibit low efficiency.

1.4 Frequency Modulation

Frequency modulation is a technique whereby the phase angle or phase shift of a carrier is varied by an applied modulating signal. The *magnitude* of the frequency change of the carrier is a direct function of the *magnitude* of the modulating signal. The *rate* at which the frequency of the carrier is changed is a direct function of the *frequency* of the modulating signal. In FM modulation, multiple pairs of sidebands are produced. The actual number of sidebands that make up the modulated wave is determined by the *modulation index* (MI) of the system.

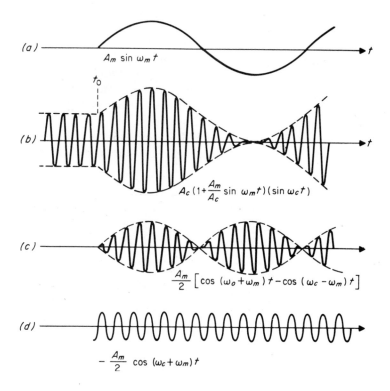

Figure 1-5 Types of suppressed-carrier amplitude modulation: (a) the modulating signal, (b) double-sideband AM signal, (c) double-sideband suppressed-carrier AM; single-sideband suppressed-carrier AM. (*Source:* Fink and Christiansen, *Electronics Engineers' Handbook, 3rd Ed.,* McGraw-Hill, New York, 1989.)

1.4.1   Modulation Index

The modulation index is a function of the frequency deviation of the system and the applied modulating signal:

$$MI = \frac{\text{Frequency deviation}}{\text{Modulating frequency}}$$

The higher the MI, the more sidebands produced. It follows that the higher the modulating frequency for a given deviation, the fewer number of sidebands produced, but the greater their spacing.

To determine the frequency spectrum of a transmitted FM waveform it is necessary to compute a Fourier series or Fourier expansion to show the actual signal components involved. This work is difficult for a waveform of this type, since the integrals that must be performed in the Fourier expansion or Fourier series are not easily solved. The result, however, is that the integral produces a particular class of solution identified as the *Bessel function* (Figure 1-6).

Supporting mathematics will show that an FM signal using the modulation indices that occur in a broadcast system will have a multitude of sidebands. From the purist point of view, *all* sidebands would have to be transmitted, received, and demodulated in order to accurately reconstruct the modulating signal. In practice, however, the channel bandwidths permitted FM systems are usually sufficient to reconstruct the modulating signal with little discernible, if not an acceptable, loss in fidelity

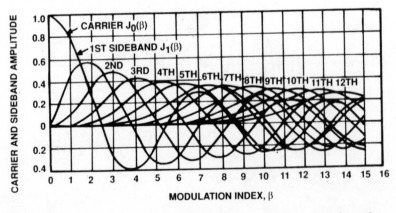

Figure 1-6 Plot of Bessel functions of the first kind as a function of argument a. (*Source:* P. F. Panter, *Modulation, Noise and Spectral Analysis*, McGraw-Hill, New York, 1965.)

FM is not a simple frequency translation, as is AM, but involves the generation of entirely new frequency components. In general, the new spectrum is much wider than the original modulating signal. This greater bandwidth may be used to improve the *signal-to-noise ratio* (S/N) of the transmission system. FM, thus, makes it possible to exchange bandwidth for S/N enhancement.

The power in an FM system is constant throughout the modulation process. While the output power is increased in an amplitude-modulation system by the modulation process, the FM system simply distributes the power throughout the various frequency components that are produced by modulation. While being modulated, a wideband FM system does not have a high amount of energy present in the carrier. Most of the energy will be found in the sum of the sidebands.

The constant-amplitude characteristic of FM greatly assists in capitalizing on the low-noise advantage of FM reception. Upon being received and amplified, the FM signal normally is clipped to eliminate all amplitude variations above a certain threshold. This removes noise picked up by the receiver as a result of man-made or atmospheric signals. It is not possible for these random noise sources to change the frequency of the desired signal; they can only affect its amplitude. The use of hard limiting in the receiver will strip off such interference.

1.4.2   Preemphasis and Deemphasis

The FM transmission/reception system offers significantly better noise rejection characteristics than AM. However, FM noise rejection is more favorable at low modulating frequencies than at high frequencies because of the reduction in the number of sidebands at higher frequencies. To offset this problem, the audio input to FM transmitters is *preemphasized* to increase the amplitude of higher-frequency signal components in normal program material. FM receivers utilize complementary *deemphasis* to produce flat overall system frequency response.

FM broadcasting uses a 75µs preemphasis curve, meaning that the time constant of the resistance-inductance (RL) or resistance-capacitance (RC) circuit used to provide the boost of high frequencies is 75 µs. Other values of preemphasis are used in different types of FM communications systems. Figure 1-7 shows three common preemphasis curves.

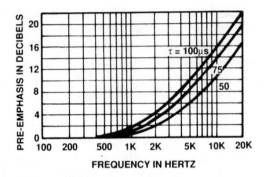

Figure 1-7 Preemphasis curves for time constants of 50, 75 and 100 μs.

1.5 Pulse Modulation

The growth of digital processing and communications has led to the development of modulation systems tailor-made for high speed, spectrum-efficient transmission. In a *pulse-modulation* system, the unmodulated carrier usually consists of a series of recurrent pulses. Information is conveyed by modulating some parameter of the pulses, such as amplitude, duration, time of occurrence, or shape. Pulse modulation is based on the *sampling principle,* which states that a message waveform with a spectrum of finite width can be recovered from a set of discrete samples if the sampling rate is higher than twice the highest sampled frequency. The samples of the input signal are used to modulate some characteristic of the carrier pulses.

1.5.1 Pulse-Amplitude Modulation

Pulse-amplitude modulation (PAM) is one of the simplest forms of data modulation. PAM departs from conventional modulation systems (described previously) in that the carrier exists as a series of pulses, rather than as a continuous waveform. The amplitude of the pulse train is modified (in accordance with the applied modulating signal) to convey intelligence (see Figure 1-8). There are two primary forms of PAM sampling:

- *Natural sampling* (or *top sampling*). The modulated pulses follow the amplitude variation of the sampled time function during the sampling interval.

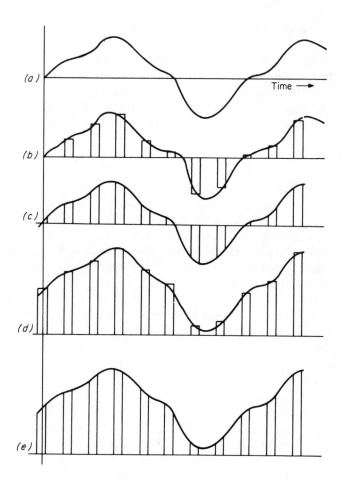

Figure 1-8 Pulse-amplitude modulation waveforms: (a) modulating signal,
(b) square-top sampling, bipolar pulse train, (c) top sampling, bipolar
pulse train, (d) square-top sampling, unipolar pulse train, (e) top sampling,
unipolar pulse train. (*Source:* Fink and Christiansen, *Electronics
Engineers' Handbook, 3rd Ed.*, McGraw-Hill, New York, 1989.)

- *Instantaneous sampling* (or *square-topped* sampling). The amplitude of the pulse is determined by the instantaneous value of the
  sampled time function corresponding to a single instant of the

sampling interval. This "single instant" may be the center or edge of the sampling interval.

There are two common methods of generating a PAM signal:

- Variation of the amplitude of a pulse sequence about a fixed non-zero value (or *pedestal*). This approach constitutes double-sideband amplitude modulation.
- Double-polarity modulated pulses with no pedestal. This approach constitutes double-sideband suppressed-carrier modulation.

1.5.2 Digital Modulation Systems

Because of the nature of digital signals (on or off), it follows that the amplitude of the signal in a pulse modulation system should be one of two heights (present, or absent/positive or negative) for maximum efficiency. Noise immunity is a significant advantage of such a system. It is necessary for the receiving system to detect only the presence or absence (or polarity) of each transmitted pulse to allow complete reconstruction of the original intelligence. The pulse shape and noise level have minimal effect (to a point). Further, if the waveform is to be transmitted over long distances, it is possible to regenerate the original signal exactly for retransmission to the next relay point. This feature is in striking contrast to analog modulation systems in which each modulation step introduces some amount of noise and signal corruption.

In any practical digital data system, some corruption of the intelligence is likely to occur over a sufficiently large span of time. Data encoding and manipulation schemes have been developed to detect and correct or conceal such errors. The addition of error correction features comes at the expense of increased system overhead, and (usually) slightly lower intelligence throughput.

Pulse-Time Modulation (PTM). A number of modulating schemes have been developed to take advantage of noise immunity afforded by a constant amplitude modulating system. *Pulse-time modulation* is one of those systems. In a PTM system, instantaneous samples of intelligence are used to vary the time of occurrence of some parameter of the pulsed carrier. Subsets of the PTM process include:

- *Pulse-duration modulation* (PDM). The time of occurrence of either the leading or trailing edge of each pulse (or both pulses) is varied from its unmodulated position by samples of the input modulating waveform. PDM may also be described as *pulse-length* or *pulse-width* modulation (PWM).

- *Pulse-position modulation* (PPM). Samples of the modulating input signal are used to vary the *position in time* of pulses, relative to the unmodulated waveform. Several types of pulse-time modulation waveforms are shown in Figure 1-9.
- *Pulse-frequency modulation* (PFM). Samples of the input signal are used to modulate the frequency of a series of carrier pulses. The PFM process is illustrated in Figure 1-10.

It should be emphasized that all of the pulse modulation systems discussed thus far may be used with both analog and digital input signals. Conversion is required for either signal into a form that can be accepted by the pulse modulator.

1.5.3   Pulse-Code Modulation

The pulse modulation systems discussed previously represent *unencoded* systems. *Pulse-code modulation* (PCM) is a scheme wherein the input signal is *quantized* into discrete steps and then sampled at regular intervals (as in conventional pulse modulation). In the *quantization* process, the input signal is sampled to produce a code representing the instantaneous value of the input within a predetermined range of values (see Figure 1-11). Only certain discrete levels are allowed in the quantization process. The code is then transmitted over the communications system as a pattern of pulses.

Quantization inherently introduces an initial error in the amplitude of the samples taken. This *quantization error* is reduced as the number of quantization steps increase. Tradeoffs must be made in system design between low quantization error, hardware complexity, and occupied bandwidth. The greater the number of quantization steps, the wider the bandwidth required to transmit the intelligence, or in the case of some signal sources, the slower the intelligence can be transmitted.

In the classic design of a PCM encoder, the quantization steps are equal. The quantization error (or *quantization noise*) can usually be reduced, however, through the use of nonuniform spacing of levels. Smaller quantization steps are provided for weaker signals, and larger steps are provided near the peak of large signals. Quantization noise is reduced by providing an encoder that is matched to the level distribution (*probability density*) of the input signal.

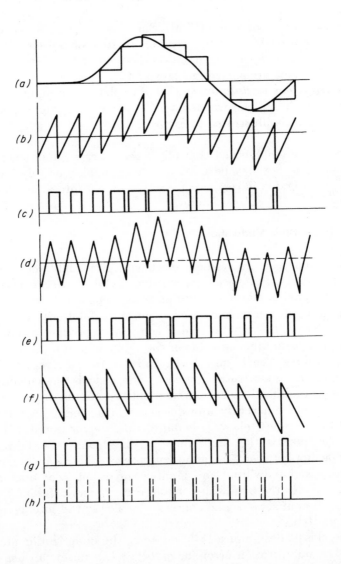

Figure 1-9 Pulse-time modulation waveforms: (a) modulating signal and sample-and-hold (S/H) waveforms, (b) sawtooth waveform added to S/H, (c) leading-edge PTM, (d) sawtooth waveform added to S/H, (e) double-edged PTM, (f) sawtooth waveform added to S/H, (g) trailing-edge PTM, (h) pulse-position modulation (reference pulse dotted). (*Source:* Fink and Christiansen, *Electronics Engineers' Handbook, 3rd Ed.,* McGraw-Hill, New York, 1989.)

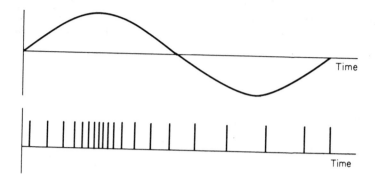

Figure 1-10 Pulse-frequency modulation. (*Source:* Fink and Christiansen, *Electronics Engineers' Handbook, 3rd Ed.,* McGraw-Hill, New York, 1989.)

Nonuniform quantization is typically realized in an encoder through processing of the input (analog) signal to compress it to match the desired nonuniformity. After compression, the signal is fed to a uniform quantization stage.

1.5.4 Delta Modulation

Delta modulation (DM) is a coding system that measures *changes* in the *direction* of the input waveform, rather than the instantaneous value of the wave itself. Figure 1-12 illustrates the concept. The clock rate is assumed to be constant. Transmitted pulses from the

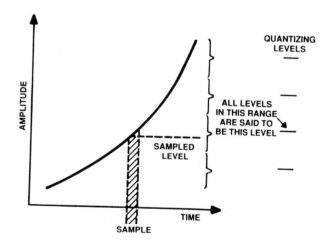

Figure 1-11 The quantization process.

pulse generator are positive if the signal is changing in a positive direction; they are negative if the signal is changing in a negative direction.

As with the PCM encoding system, quantization noise is a parameter of concern for DM. Quantization noise can be reduced by increasing the sampling frequency (the pulse generator frequency). The DM system has no fixed maximum (or minimum) signal amplitude. The limiting factor is the slope of the sampled signal, which must not change by more than one level or step during each pulse interval.

1.5.5   Digital Coding Systems

A number of methods exist to transmit digital signals over long distances in analog transmission channels. Some of the more common systems include:

- *Binary on-off keying* (BOOK). A method by which a high-frequency sinusoidal signal is switched on and off corresponding to 1 and 0 (on and off) periods in the input digital data stream. In practice, the transmitted sinusoid waveform does not start or stop abruptly, but follows a predefined ramp up or down.
- *Binary frequency-shift keying* (BFSK). A modulation method in which a continuous wave is transmitted that is shifted between two frequencies, representing 1s and 0s in the input data stream. The BFSK signal may be generated by switching between two oscillators (set to different operating frequencies), or by applying a binary baseband signal to the input of a *voltage-controlled oscillator* (VCO). The transmitted signals are often referred to as a *mark*

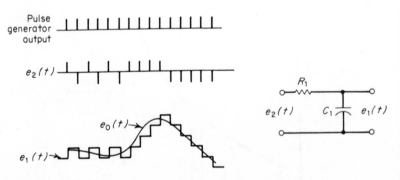

Figure 1-12 Delta modulation waveforms using single integration. (*Source:* Fink and Christiansen, *Electronics Engineers' Handbook, 3rd Ed.,* McGraw-Hill, New York, 1989.)

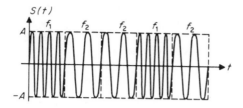

Figure 1-13 Binary FSK waveform. (*Source:* Fink and Christiansen, *Electronics Engineers' Handbook, 3rd Ed.,* McGraw-Hill, New York, 1989.)

(binary digit 1) or a *space* (binary digit 0). Figure 1-13 illustrates the transmitted waveform of a BFSK system.

- *Binary phase-shift keying* (BPSK). A modulating method in which the phase of the transmitted wave is shifted 180° in synchronism with the input digital signal. The phase of the RF carrier is shifted by $+\pi/2$ radians or $-\pi/2$ radians, depending upon whether the data bit is a 0 or a 1. Figure 1-14 shows the BPSK transmitted waveform.
- *Quadriphase shift keying* (QPSK). A modulation scheme similar to BPSK except that quaternary modulation is employed, rather than binary modulation. QPSK requires half the bandwidth of BPSK for the same transmitted data rate.

1.5.6   Spread Spectrum Systems

The specialized requirements of the military have led to the development of *spread-spectrum* communications systems. As the name implies, such systems require a frequency range substantially greater

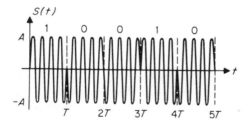

Figure 1-14 Binary PSK waveform. (*Source:* Fink and Christiansen, *Electronics Engineers' Handbook, 3rd Ed.,* McGraw-Hill, New York, 1989.)

than the basic information-bearing signal. Spread-spectrum systems have some or all of the following properties:

- Low interference to other communications systems
- Ability to reject high levels of external interference
- Immunity to jamming by hostile forces
- Provides for secure communications paths
- Operates over multiple RF paths

Spread-spectrum systems operate within an entirely different set of requirements than transmission systems discussed previously. Conventional modulation methods are designed to provide for the easiest possible reception and demodulation of the transmitted intelligence. The goals of spread-spectrum systems, on the other hand, are secure and reliable communications that cannot be intercepted by unauthorized persons. The most common modulation and encoding techniques in spread-spectrum communications include:

- *Frequency hopping.* A random or *pseudorandom number* (PN) sequence is used to change the carrier frequency of the transmitter. This approach has two basic variations: *slow frequency hopping*, where the hopping rate is smaller than the data rate; and *fast frequency hopping*, where the hopping rate is larger than the data rate. In a fast frequency hopping system, the transmission of a single piece of data occupies more than one frequency. Frequency hopping systems permit multiple-access capability to a given band of frequencies because each transmitted signal occupies only a fraction of the total transmitted bandwidth.
- *Time hopping.* A PN sequence is used to switch the position of a message-carrying pulse within a series of frames.
- *Message corruption.* A PN sequence is added to the message before modulation.
- *Chirp spread spectrum.* Linear frequency modulation of the main carrier is used to spread the transmitted spectrum. This technique is commonly used in radar and has also been applied to communications systems.

In a spread-spectrum system, the signal power is divided over a large bandwidth. The signal, therefore, has a small average power in any single narrow-band slot. This means that a spread-spectrum system can share a given frequency band with one or more narrow-band systems. Further, because of the low energy in any particular band, detection or interception of the transmission is difficult. Figure 1-15 summarizes the primary methods of spread-spectrum transmission.

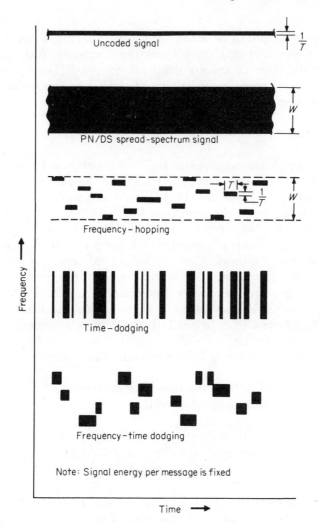

Figure 1-15 Spectral characteristics of spread-spectrum transmission. (*Source:* Fink and Christiansen, *Electronics Engineers' Handbook, 3rd Ed.,* McGraw-Hill, New York, 1989.)

detection or interception of the transmission is difficult. Figure 1-15 summarizes the primary methods of spread-spectrum transmission.

1.6   RF Power Amplifiers

The process of generating high power RF signals has been refined over the years to an exact science. Advancements in devices and circuit design continue to be made each year, pushing ahead the barriers of efficiency and maximum operating frequency. Although different applications place unique demands on the RF design engineer, the fundamental concepts of RF amplification are applicable to virtually any system.

1.6.1   Frequency Source

Every RF amplifier requires a stable frequency reference. At the heart of most systems is a quartz crystal. Quartz acts as a stable high Q mechanical resonator. Crystal resonators are available for operation at frequencies ranging from 1 kHz to 300 MHz. The operating characteristics of a crystal are determined by the *cut* of the device from a bulk "mother" crystal. The behavior of the device depends strongly on the size and shape of the crystal, and the angle of the cut. To provide for operation at a wide range of frequencies, different cuts, vibrating in one or more selected modes are used. Crystals are temperature-sensitive, as shown in Figure 1-16. The extent to which a device is affected by changes in temperature is determined by its cut and packaging. Crystals also exhibit changes in frequency with time. Such *aging* is caused by one or both of the following:

• Mass transfer to or from the resonator surface
• Stress relief within the device itself

Crystal aging is most pronounced when the device is new. As stress within the internal structure is relieved, the aging process slows.

Frequency Stabilization. The stability of a quartz crystal is inadequate for most commercial and industrial applications. Two common methods are used to provide the required long-term frequency stability:

• *Oven-controlled crystal oscillator.* A technique where the crystal is installed in a temperature-controlled box. Because the temperature is constant in the box, controlled by a thermostat, the crystal remains on-frequency. The temperature of the enclosure is usually set to the *turnover temperature* of the crystal.
• *Temperature-compensated crystal oscillator* (TCXO). A technique where the frequency-vs.-temperature changes of the crystal are compensated by varying a load capacitor. A thermistor network is

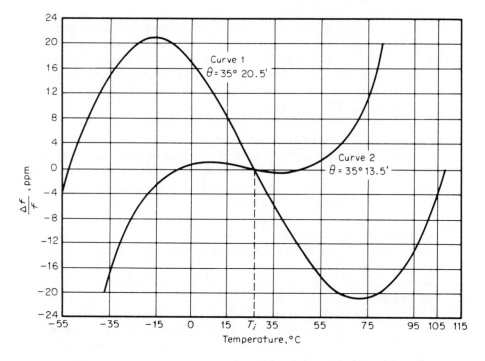

Figure 1-16 The effects of temperature on two types of AT-cut crystals. (*Source:* Fink and Christiansen, *Electronics Engineers' Handbook, 3rd Ed.,* McGraw-Hill, New York, 1989.)

typically used to generate a correction voltage that feeds a varactor to retune the crystal to the desired on-frequency value.

1.6.2   Operating Class

Power amplifier (PA) stage operating efficiency is a key element in the design and application of an RF system. As the power level of an RF generator increases, the overall efficiency of the system becomes more important. Increased efficiency translates into lower operating costs and usually improved reliability of the system. The operating mode of the final stage, or stages, is the primary determining element in the maximum possible efficiency of the system.

All electron amplifying devices are classified by their *class of operation.* Four primary class divisions apply to RF generators:

- *Class A.* A mode wherein the power amplifying device is operated over its linear transfer characteristic. This mode provides the lowest waveform distortion, but also the lowest efficiency. The basic operating efficiency of a class A stage is 50 percent. Class A amplifiers exhibit low intermodulation distortion, making them well suited to linear RF amplifier applications.
- *Class B.* A mode wherein the power amplifying device is operated just outside its linear transfer characteristic. This mode provides improved efficiency at the expense of some waveform distortion. Class AB is a variation on class B operation. The transfer characteristic for an amplifying device operating in this mode is, predictably, between class A and class B.
- *Class C.* A mode wherein the power amplifying device is operated significantly outside its linear transfer characteristic, resulting is a pulsed output waveform. High efficiency (up to 90 percent) can be realized with class C operation; however, significant distortion of the waveform will occur. Class C is used exclusively as an efficient RF power generator.
- *Class D.* A mode that essentially results in a switched device state. The power amplifying device is either *on* or *off*. This is the most efficient mode of operation. It is also the mode that produces the greatest waveform distortion.

The angle of current flow determines the class of operation for a power amplifying device. Typically, the conduction angle for class A is 360°; class AB is between 180° and 360°; class B is 180°; and class C is less than 180°. Subscripts may also be used to denote grid current flow, in the case of a power vacuum tube device. The subscript "1" means that no grid current flows in the stage; the subscript "2" denotes grid current flow. Figure 1-17 charts operating efficiency as a function of the conduction angle for an RF amplifier.

The class of operation is not directly related to the type of amplifying circuit. Vacuum tube stages may be grid- or cathode-driven without regard to the operating class. Similarly, solid state amplifiers may be configured for grounded emitter, grounded base, or grounded collector operation without regard to the class of operation.

Operating Efficiency. The design goal of all RF amplifiers is to convert input power into an RF signal at the greatest possible efficiency. DC input power that is not converted to a useful output signal is, for the most part, converted to heat. This heat represents wasted energy, which must be removed from the amplifying device. Removal of heat is a problem common to all high power RF amplifiers. Cooling methods include:

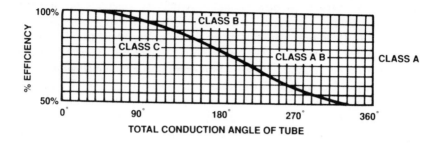

Figure 1-17 Plate efficiency as a function of conduction angle for an amplifier with a tuned load.

- Natural convection
- Radiation
- Forced convection
- Liquid
- Conduction
- Evaporation

The type of cooling method chosen is dictated, in large part, by the type of active device used and the power level involved. For example, liquid cooling is used almost exclusively for high power (100 kW) vacuum tubes; conduction is used most often for low power (20 W) transistors.

1.6.3   Broadband Amplifier Design

RF design engineers face a continuing challenge to provide adequate bandwidth for the signals to be transmitted, while preserving as much efficiency from the overall system as possible. These two parameters, while not mutually exclusive, often involve tradeoffs for both designers and operators.

An ideal RF amplifier will operate over a wide band of frequencies with minimum variations in output power, phase, distortion, and efficiency. The bandwidth of the amplifier depends to a great extent on the type of active device used, the frequency range required, and the operating power. As a general rule, bandwidth of 20 percent or greater at frequencies above 100 MHz can be considered *broadband*. Below 100 MHz, broadband amplifiers typically have a bandwidth of one octave or more.

Most development in new broadband designs focuses on semiconductor technology. Transistor and MOSFET (metal oxide semiconductor field effect transistor) devices have ushered in the era of *distributed amplification,* where multiple devices are used to achieve the required RF output power. Semiconductor-based designs offer benefits beyond active device redundancy. Bandwidth at frequencies above 100 MHz can often be improved because of the smaller physical size of semiconductor devices, which translates into reduced lead and component inductance and capacitance.

Amplifier Compensation. A variety of methods may be used to extend the operating bandwidth of a transistorized amplifier stage. Two of the most common methods are series- and shunt-compensation circuits, shown in Figure 1-18. These two basic techniques can be combined, as shown. Other circuit configurations may be used for specific requirements, such as phase compensation.

Stagger Tuning. Several stages with narrowband response (relative to the desired system bandwidth) can be cascaded and, through the use of *stagger tuning,* made broadband. While there is an efficiency penalty for this approach, it has been used for years in all types of equipment. The concept is simple: offset the center operating frequencies (and, therefore, peak amplitude response) of the cascaded amplifiers so the resulting passband is flat and broad.

For example, the first stage in a three-stage amplifier is adjusted for peak response at the center operating frequency of the system. The second stage is adjusted above the center frequency, and the third stage is adjusted below the center. The resulting composite response curve yields a broadband trace. The efficiency penalty for this scheme varies depending on the power level of each stage, the amount of stagger tuning required to achieve the desired bandwidth, and the number of individual stages.

Matching Circuits. The individual stages of an RF generator must be coupled together. Rarely does the output impedance and power level of one stage precisely match the input impedance and signal-handing level of the next stage. There is a requirement, therefore, for broadband matching circuits. Matching at RF frequencies can be accomplished with several different techniques, including:

- *Quarter-wave transformer.* A matching technique using simply a length of transmission line 1/4-wave long, with a characteristic impedance of:

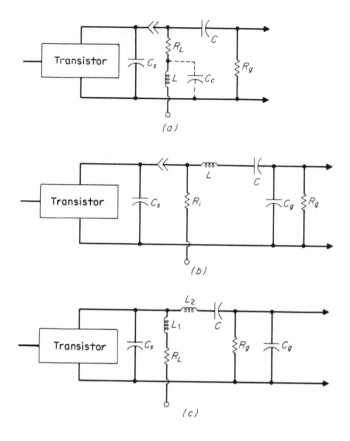

Figure 1-18 High frequency compensation techniques: (a) shunt, (b) series, (c) combination of shunt and series. (*Source:* Fink and Christiansen, *Electronics Engineers' Handbook, 3rd Ed.,* McGraw-Hill, New York, 1989.)

$$Z_{line} = \sqrt{Z_{in} \times Z_{out}}$$

Where:
Z_{in} and Z_{out} = the terminating impedances

Quarter-wave transformers can be cascaded to achieve more favorable matching characteristics. Cascaded transformers permit small matching ratios for each individual section.

- *Balun transformer.* A transmission-line transformer in which the turns are arranged physically to include the interwinding capacitance as a component of the characteristic impedance of the transmission line. This technique permits wide bandwidths to be achieved without unwanted resonances. Balun transformers usually are made of twisted wire pairs or twisted coaxial lines. Ferrite toroids may be used as the core material.

- Other types of lumped reactances.
- Short sections of transmission line.

Power Combining. The two most common methods of extending the operating power of semiconductor devices are direct paralleling of components and *hybrid splitting/combining*. Direct paralleling has been used for both tube and semiconductor designs; however, application of this simple approach is limited by variations in device operating parameters. Two identical devices in parallel do not necessarily draw the same amount of current (supply the same amount of power to the load). Paralleling at UHF frequencies and above can be difficult because of the restrictions of operating wavelength.

The preferred approach involves the use of identical stages driven in parallel from a *hybrid coupler*. The coupler provides a constant-source impedance, and directs any reflected energy from the driven stages to a *reject port* for dissipation. A hybrid coupler offers a VSWR-canceling effect that improves system performance. Hybrids also provide a high degree of isolation between active devices in a system.

1.6.4   Output Devices

Significant changes have occurred within the past five years or so with regard to power amplifying devices. Vacuum tubes were the mainstay of radio frequency transmission equipment until advanced semiconductor components became available at competitive prices. Many high power applications that demanded vacuum tubes can now be met with solid state devices arranged in a distributed amplification system. MOSFET and bipolar components have been used successfully in radio and television broadcast transmitters, shortwave transmitters, sonar transmitters, induction heaters, and countless other applications.

Most solid state designs used today are not simply silicon versions of classic vacuum tube circuits. They are designed to maximize efficiency through class D switching, and maximize reliability through distributed amplification and redundancy.

The principal drawback of a solid state system over a vacuum tube design of comparable power is the circuit complexity that goes with most semiconductor-based hardware. Preventive maintenance is reduced significantly and, in theory, repair is simpler as well in a solid state system. The parts count in almost all semiconductor-based hardware, however, is significantly greater than in a comparable tube system. Increased parts translate to (usually) a higher initial purchase price for the equipment and increased vulnerability to device failure.

Efficiency comparisons between vacuum tube and solid state systems do not always yield the dramatic contrasts expected. While most semiconductor amplifiers incorporate switching technology that is far superior to class B or C operation (not to mention class A), power losses are experienced in the signal splitting and combining networks necessary to make distributed amplification work.

It is evident, then, that vacuum tubes and semiconductors each have their benefits and drawbacks. Both technologies will remain viable for many years to come. Vacuum tubes will not go away, but will move to higher power levels and higher operating frequencies.

Bibliography

1. Fink, D. and D. Christiansen, *Electronics Engineer's Handbook*, Third Edition, McGraw-Hill, New York, NY, 1989.
2. Jordan, Edward C., *Reference Data for Engineers: Radio, Electronics, Computer and Communications*, Seventh Edition, Howard W. Sams Company, Indianapolis, IN, 1985.
3. Hulick, Timothy P., Using Tetrodes for High Power UHF, *Proceedings of the Society of Broadcast Engineers National Convention*, vol. 4, pages 52–57, SBE, Indianapolis, IN, 1989.
4. The Laboratory Staff, *The Care and Feeding of Power Grid Tubes*, Varian Eimac, San Carlos, CA, 1982.

2

Broadcast Applications

2.1 Introduction

Broadcasting has been around for a long time. Amplitude modulation (AM) was the first modulation system that permitted voice communications to take place. This simple modulation system was predominant through the 1920s and 1930s. Frequency modulation (FM) came into regular broadcast service during the 1940s. Television broadcasting, which uses amplitude modulation for the visual portion of the signal and frequency modulation for the aural portion of the signal, became available to the public during the mid-1940s. These two simple approaches to modulating a carrier have served the broadcast industry well for many decades. While these basic modulating schemes still exist today, numerous enhancements have been made, including: color, stereo operation, and subcarrier programming. New technology has given broadcasters new ways to serve the public.

Technology has also changed the rules by which the broadcasting systems developed. AM radio, as a technical system, offered limited audio fidelity, but provided design engineers with a system that allowed uncomplicated transmitters and simple, inexpensive receivers. FM radio, on the other hand, offered excellent audio fidelity, but required a complex and unstable transmitter (in the early days) and complex, expensive receivers. It is, therefore, no wonder that AM radio flourished and FM remained relatively stagnant for at least 20 years after being introduced to the public.

Television broadcasting evolved slowly during the late 1940s and early '50s. Color transmissions were authorized as early as 1952.

Color receivers were not purchased by consumers in large numbers, however, until the mid-1960s. Early color sets suffered from poor reliability, unstable performance, and high cost to consumers. All that changed with the introduction of transistors, and later, integrated circuits (ICs). Most color TV receivers produced today consist of an IC chip set numbering from 8 to 15 devices.

2.2   AM Radio Broadcasting

AM radio stations operate on 10 kHz channels spaced evenly from 540 kHz to 1600 kHz. Various classes of stations have been established by the Federal Communications Commission (FCC), as well as agencies in other countries, to allocate the available spectrum to given regions and communities. In the U.S., the basic classes are *clear, regional,* and *local.* Current practice uses the CCIR (international) designations as class A, B and C, respectively. Operating power levels range from 50 kW for a clear channel station to as little as 250 W for a local station.

AM stations choosing to do so may operate in stereo using one of two incompatible formats. Known in the broadcast industry as the *Kahn system* and *CQUAM system,* both formats are fully compatible with existing monophonic radios. (CQUAM is a registered trademark of Motorola, Inc.) In order to receive a stereo AM broadcast, consumers must purchase a new stereo radio. Depending on the design of the radio, however, the consumer may or may not be able to receive in stereo two stereo AM stations in the same area transmitting with different systems. The radio, instead, usually will receive one system in stereo, and the other station (using the competing system) in mono.

This unusual situation is the result of a decision by the FCC to let the "marketplace" decide which system, Kahn or CQUAM, would be the defacto broadcast standard. Unfortunately, the marketplace has not, as of this writing, completely decided which way to go. This confusion has slowed the introduction of stereo AM broadcasting and the purchase by consumers of new, high-fidelity stereo AM receivers.

The two systems have fundamental differences. The Kahn system independently modulates the sideband signals of the AM carrier, one for the right channel and the other for the left channel, to provide the stereo effect. The CQUAM system transmits the stereo *sum signal* (in other words, the monophonic signal) in the usual manner, and places the stereo *difference signal* on a phase-modulate subchannel. Decoder circuits in the receiver reconstruct the stereo signals.

2.2.1   High Level AM Modulation

High level anode modulation is the oldest and simplest way of generating a high power AM signal. In this system, the modulating signal is amplified and combined with the dc supply source to the anode of the final RF amplifier stage. The RF amplifier is normally operated class C. The final stage of the modulator usually consists of a pair of tubes operating class B in a push-pull configuration. A basic high-level modulator is shown in Figure 2-1.

The RF signal is normally generated in a low-level transistorized oscillator. It is then amplified by one or more solid state or vacuum tube stages to provide final RF drive at the appropriate frequency to the grid of the final class C amplifier. The audio input is applied to an intermediate power amplifier (usually solid state) and used to drive two class B (or class AB) push-pull output devices. The final amplifiers provide the necessary modulating power to drive the final RF stage. For 100 percent modulation, this modulating power is equal to 50 percent of the actual carrier power.

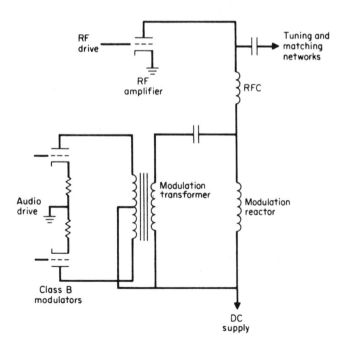

Figure 2-1 Simplified diagram of a high-level amplitude-modulated amplifier. (*Source:* K. Blair Benson, *Audio Engineering Handbook,* McGraw-Hill, New York, 1988.)

The modulation transformer shown in the figure does not usually carry the dc supply current for the final RF amplifier. The modulation reactor and capacitor shown provide a means to combine the audio signal voltage from the modulator with the dc supply to the final RF amplifier. This arrangement eliminates the necessity of having dc current flow through the secondary of the modulation transformer, which would result in magnetic losses and saturation effects. In some newer transmitter designs, the modulation reactor has been eliminated from the system, thanks to improvements in transformer technology.

The RF amplifier normally operates class C with grid current drawn during positive peaks of the cycle. Typical stage efficiency is 75 percent to 83 percent. An RF tank following the amplifier resonates the output signal at the operating frequency and, with the assistance of a low pass filter, eliminates harmonics of the amplifier caused by class C operation.

This type of system was popular in AM broadcasting for many years, primarily because of its simplicity. The primary drawback is overall low system efficiency. The class B modulator tubes cannot operate with greater than 50 percent efficiency. Still, with inexpensive electricity, this was not considered to be a significant problem. As energy costs increased, however, more efficient methods of generating high power AM signals were developed. Increased efficiency normally came at the expense of added technical complexity.

2.2.2   Pulse-Width Modulation

Pulse-width modulation (PDM), also known as *pulse-duration modulation* is one of the most popular systems developed for modern vacuum tube AM transmitters. Figure 2-2 shows the PDM scheme as patented by the Harris Corporation. The PDM system works by utilizing a square wave switching system, illustrated in Figure 2-3.

The PDM process begins with a signal generator (see Figure 2-4). A 75 kHz sine wave is produced by an oscillator and used to drive a square wave generator, resulting in a simple 75 kHz square wave. The square wave is then integrated, resulting in a triangular waveform that is mixed with the input audio in a summing circuit. The resulting signal is a triangular waveform that rides on the incoming audio. This composite signal is then applied to a threshold amplifier, which functions as a switch, that is turned on whenever the value of the input signal exceeds a certain limit. The result is a string of pulses in which the width of the pulse is proportional to the period of time the triangular waveform exceeds the threshold. The pulse out-

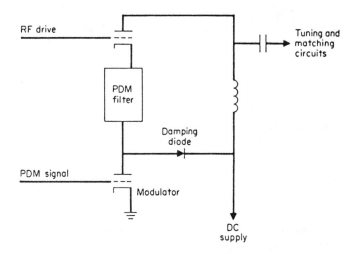

Figure 2-2 The pulse-duration modulation (PDM) method of pulse-width modulation (patented by Harris Corp.). (*Source:* K. Blair Benson, *Audio Engineering Handbook,* McGraw-Hill, New York, 1988.)

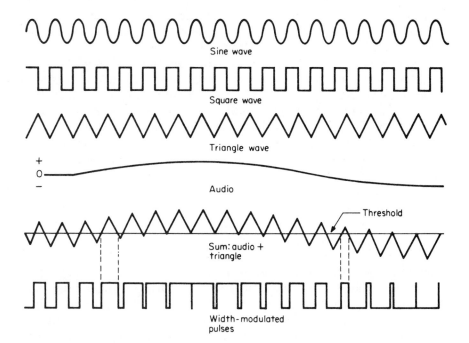

Figure 2-3 The principal waveforms of the PDM system. (*Source:* K. Blair Benson, *Audio Engineering Handbook,* McGraw-Hill, New York, 1988.)

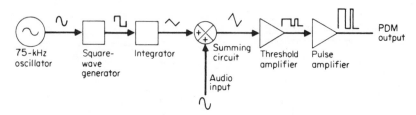

Figure 2-4 Block diagram of a PDM waveform generator. (*Source:* K. Blair Benson, *Audio Engineering Handbook,* McGraw-Hill, New York, 1988.)

put is applied to an amplifier to obtain the necessary power to drive subsequent stages. A filter eliminates whatever transients may exist after the switching process is complete.

The PDM scheme is, in effect, a digital modulation system with the audio information being sampled at a 75 kHz rate. The width of the pulses contains all the audio information. The pulse-width-modulated signal is applied to a *switch* or *modulator tube*. The tube is simply turned *on,* to a fully saturated state, or *off* in accordance with the instantaneous value of the pulse. When the pulse goes positive, the modulator tube is turned on and the voltage across the tube drops to a minimum. When the pulse returns to its minimum value, the modulator tube turns off.

This PDM signal becomes the power supply to the final RF amplifier tube. When the modulator is switched on, the final amplifier will experience current flow and RF will be generated. When the switch or modulator tube goes off, the final amplifier current will cease. This system causes the final amplifier to operate in a highly efficient class D switching mode. A dc offset voltage to the summing amplifier is used to set the carrier (no modulation) level of the transmitter.

A high degree of third-harmonic energy will exist at the output of the final amplifier because of the switching-mode operation. This energy is eliminated by a third-harmonic trap. The result is a stable amplifier that normally operates in excess of 90 percent efficiency. The power consumed by the modulator and its driver is usually a fraction of a full class B amplifier stage.

The *damping diode,* shown in the previous figure, is included to prevent potentially damaging transient over-voltages during the switching process. When the switching tube turns off the supply current during a period where the final amplifier is conducting, the high current carried through the inductors contained in the PDM filters could cause a large transient voltage to be generated. The energy in the PDM filter is returned to the power supply by the damping diode. If no alternative route were established, the energy would return by arcing through the modulator tube itself.

The pulse-width-modulation system makes it possible to completely eliminate audio frequency transformers in the transmitter. The result is wide frequency response and low distortion. It should be noted that variations on this amplifier and modulation scheme have been used by other manufacturers for both standard broadcast and shortwave service.

2.2.3 Digital Modulation

Current transmitter design work for AM broadcasting has focused almost exclusively on solid state technology. High power MOSFET devices and digital modulation techniques have made possible a new generation of energy-efficient systems, with audio performance that easily surpasses vacuum tube designs. Most solid state AM systems operate in a highly efficient class D switching mode. Multiple MOSFET driver boards are combined through one of several methods to achieve the required carrier power. Figure 2-5 shows one such system.

A low power RF drive signal at the carrier frequency is applied through a multi-secondary transformer to FET pairs Q1 and Q3. The FETs switch on and off at the carrier frequency, producing a square wave of reversing polarity at the primary of T2. The output signal is

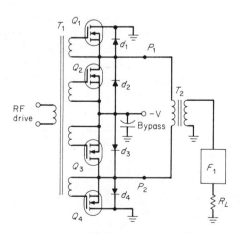

Figure 2-5 Basic design of a solid state (Nautel) switching RF power amplifier designed for MF operation. (*Source:* Fink and Christiansen, *Electronics Engineers' Handbook, 3rd Ed.,* McGraw-Hill, New York, 1989.)

applied to a low pass filter to remove the harmonics of the carrier, and then to the load. Stage efficiency of 90 percent or greater is possible with this design.

2.2.4   Shortwave Broadcasting

The technologies used in commercial and government-sponsored shortwave broadcasting are closely allied with those used in AM radio. Shortwave stations usually, however, operate at significantly higher powers than AM stations.

International broadcast stations use frequencies ranging from 5.95 MHz to 26.1 MHz. These transmissions are intended for reception by the general public in foreign countries. Figure 2-6 shows the frequencies assigned by the FCC for international broadcast shortwave service. The minimum output power is 50 kW. Assignments are made for specific hours of operation at specific frequencies.

Very high power shortwave transmitters have been installed to serve large geographical areas, and to overcome jamming efforts by foreign governments. Systems rated for power outputs of 500 kW and more are not uncommon. RF circuits designed specifically for high power operation are utilized.

Most shortwave transmitters have the unique requirement of automatic tuning to one of several preset operating frequencies. A variety of schemes exist to accomplish this task, including multiple exciters (each set to the desired operating frequency) and motor-controlled variable inductors and capacitors. Tune-up at each frequency is made by the transmitter manufacturer. The settings of all tuning controls are stored in a memory device, or as a set of trim potentiometer adjustments. Automatic retuning of a high power shortwave transmitter can be accomplished in less than 30 seconds in most cases.

Band	Frequency, kHz	Meter band, m
A	5,950–6,200	49
B	9,500–9,775	32
C	11,700–11,975	25
D	15,100–15,450	19
E	17,700–17,900	16
F	21,450–21,750	14
G	25,600–26,100	11

Figure 2-6 Operating frequency bands for shortwave broadcasting. (*Source:* Fink and Christiansen, *Electronics Engineers' Handbook, 3rd Ed.,* McGraw-Hill, New York, 1989.)

Power Amplifier Types. Shortwave technology has advanced significantly within the last five years, thanks to improved semiconductor devices. High power MOSFETs and other components have made solid state shortwave transmitters, operating at 100 kW and higher, practical. The vast majority of shortwave systems now in use, however, use vacuum tubes as the power-generating element.

The efficiency of a power amplifier/modulator for shortwave applications is of critical importance. Because of the power levels involved, low efficiency translates into higher operating costs. A number of transmitters currently in use rely on the pulse duration modulation system described previously to achieve good operating efficiency. The PDM system permits easy scaling of the power control components necessary to meet the required RF output.

Older, more traditional tube-type shortwave transmitters typically utilize one of the following modulation systems:

• Doherty amplifier
• Chireix outphasing modulated amplifier
• Dome modulated amplifier
• Terman-Woodyard modulated amplifier

The operation of these systems is described in Chapter 7.

2.3 FM Radio Broadcasting

FM radio stations operate on 200 kHz channels spaced evenly from 88.1 MHz to 107.9 MHz. In the U.S., channels below 92.1 MHz are reserved for non-commercial, educational stations. The FCC has established three classifications for FM stations operating east of the Mississippi River and four classifications for stations west of the Mississippi. Power levels range from a high of 100 kW *effective radiated power* (ERP) to 3 kW or less for a lower classification. The ERP of a station is a function of transmitter power output (TPO) and antenna gain. ERP is determined by multiplying these two quantities together and allowing for line loss.

A transmitting antenna is said to have "gain" if, by design, it concentrates useful energy at low radiation angles, rather than allowing a substantial amount of energy to be radiated above the horizon (and be lost in space). FM and TV transmitting antennas are designed to provide gain by stacking individual radiating elements vertically. At first examination, it might seem reasonable and economical to achieve licensed ERP using the lowest transmitter power output possible and highest antenna gain. Other factors, however, come into play that complicate the situation. Factors that limit the use of high gain antennas include:

- Effects of high gain designs on coverage area and signal penetration
- Limitations on antenna size because of tower restrictions, such as available vertical space, weight and windloading
- Cost of the antenna

Stereo broadcasting is used almost universally in FM radio today. Introduced in the mid-1960s, stereo has contributed in large part to the success of FM radio. The left and right sum (monophonic) information is transmitted as a standard frequency-modulated signal. Filters restrict this *main channel* signal to a maximum of about 17 kHz. A pilot signal is transmitted at low amplitude at 19 kHz to enable decoding at the receiver. The left and right difference signal is transmitted as an amplitude-modulated subcarrier that frequency-modulates the main FM carrier. The center frequency of the subcarrier is 38 kHz. Decoder circuits in the FM receiver matrix the sum and difference signals to reproduce the left and right audio channels. Figure 2-7 illustrates the baseband signal of a stereo FM station.

2.3.1   Modulation Circuits

Early FM transmitters used *reactance modulators* that operated at a low frequency. The output of the modulator was then multiplied to

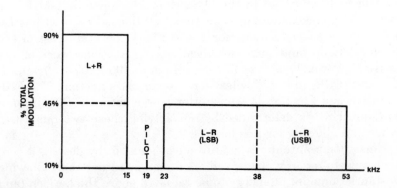

Figure 2-7 Composite baseband stereo FM signal. A full left-only or right-only signal will modulate the main (L+R) channel to a maximum of 45 percent. The stereophonic subchannel is composed of upper sideband (USB) and lower sideband (LSB) components. (*Source:* Benson and Whitaker, *Television and Audio Handbook,* McGraw-Hill, New York, 1990.)

reach the desired output frequency. This approach was acceptable for monaural FM transmission, but not for modern stereo systems or other applications that utilize subcarriers on the FM broadcast signal. Modern FM systems all utilize what is referred to as *direct modulation*. That is, frequency modulation occurs in a modulated oscillator that operates on a center frequency equal to that of the desired transmitter output frequency. In stereo broadcast systems, a composite FM signal is applied to the FM modulator.

Various techniques have been developed to generate the direct-FM signal. One of the most popular techniques uses a variable-capacity diode as the reactive element in the oscillator. The modulating signal is applied to the diode, which causes the capacitance of the device to vary as a function of the magnitude of the modulating signal. Variations in the capacitance cause the frequency of the oscillator to vary. Again, the magnitude of the frequency shift is proportional to the amplitude of the modulating signal, and the rate of frequency shift is equal to the frequency of the modulating signal.

The direct-FM modulator is one element of an FM transmitter exciter, which generates the composite FM waveform. A block diagram of a complete FM exciter is shown in Figure 2-8. Audio inputs of various types (stereo left and right signals, plus subcarrier programming, if used) are buffered, filtered, and preemphasized before being *summed* to feed the modulated oscillator. It should be noted that the oscillator is not normally coupled directly to a crystal, but to a free-running oscillator adjusted as closely as possible to the carrier frequency of the transmitter. The final operating frequency is carefully maintained by an automatic frequency control system employing a *phase locked loop* (PLL) tied to a reference crystal oscillator or frequency synthesizer. (The operation of a PLL is described in Section 2.4.5.)

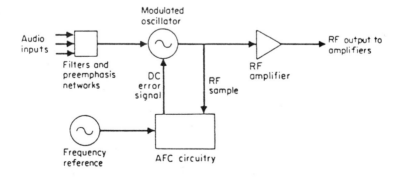

Figure 2-8 Block diagram of a FM exciter. (*Source:* K. Blair Benson, *Audio Engineering Handbook,* McGraw-Hill, New York, 1988.)

A solid-state class C amplifier follows the modulated oscillator and raises the operating power of the FM signal to 20-30 W. One or more subsequent amplifiers in the transmitter raise the signal power to several hundred watts for application to the final power amplifier stage. Nearly all current high power FM transmitters utilize solid-state amplifiers up to the final RF stage, which is generally a vacuum tube for operating powers of 5 kW and above. All stages operate in the class C mode. In contrast to AM systems, each stage in an FM power amplifier can operate in class C because no information is lost from the frequency-modulated signal as a result of amplitude changes. As mentioned previously, FM is a constant-power system.

Auxiliary Services. Modern FM broadcast stations are capable of not only broadcasting stereo programming, but one or more subsidiary channels as well. These signals, referred to by the FCC as *Subsidiary Communications Authorization* (SCA) services, are used for the transmission of stock market data, background music, control signals, and other information not normally part of the station's main programming. These services do not provide the same range of coverage or audio fidelity as the main stereo program; however, they perform a public service and can represent a valuable source of income for the broadcaster.

SCA systems provide efficient use of the available spectrum. The most common subcarrier frequency is 67 kHz, although higher subcarrier frequencies may be utilized. Stations that operate subcarrier systems are permitted by the FCC to exceed (by a small amount) the maximum 75 kHz deviation limit, under certain conditions. The subcarriers utilize low modulation levels, and the energy produced is maintained essentially within the 200 kHz bandwidth limitation of FM channel radiation.

2.3.2   FM Power Amplifiers

Nearly all high power FM transmitters manufactured today employ cavity designs. The 1/4-wavelength cavity is the most common. The design is simple and straightforward. A number of variations can be found in different transmitters, but the underlying theory of operation is the same. The goal of any cavity amplifier is to simulate a resonant tank circuit at the operating frequency, and provide a means of coupling the energy in the cavity to the transmission line. Because of the operating frequencies involved (88 MHz to 108 MHz) the elements of the "tank" take on unfamiliar forms.

A typical 1/4-wave cavity is shown in Figure 2-9. The plate of the tube connects directly to the inner section (tube) of the plate-blocking capacitor. The blocking capacitor can be formed in one of several ways. In at least one design, it is made by wrapping the outside surface of the inner tube conductor with multiple layers of insulating film. The exhaust chimney/inner conductor forms the other element of the blocking capacitor. The cavity walls form the outer conductor of the 1/4-wave transmission line circuit. The dc plate voltage is applied to the PA tube by a cable routed inside the exhaust chimney and inner tube conductor. In this design, the screen-contact finger-stock ring mounts on a metal plate that is insulated from the grounded-cavity deck by a blocking capacitor. This hardware makes up the screen-blocker assembly. The dc screen voltage feeds to the fingerstock ring from underneath the cavity deck through an insulated feedthrough.

Some transmitters employing the 1/4-wave cavity design use a grounded-screen configuration in which the screen contact finger-stock ring is connected directly to the grounded cavity deck. The PA cathode then operates at below ground potential (in other words, at a

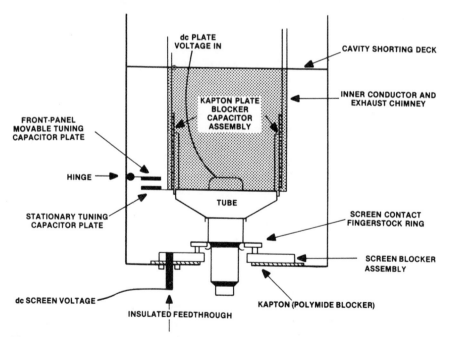

Figure 2-9 The layout of a common type of 1/4-wave PA cavity for FM broadcast service. (*Source: Broadcast Engineering Magazine.*)

negative voltage), establishing the required screen voltage for the tube.

Coarse tuning of the cavity is accomplished by adjusting the cavity length. The top of the cavity (the cavity shorting deck) is fastened by screws or clamps and can be raised or lowered to set the length of assembly for the particular operating frequency. Fine tuning is accomplished by a variable-capacity plate-tuning control built into the cavity. In the example, one plate of this capacitor, the stationary plate, is fastened to the inner conductor just above the plate-blocking capacitor. The movable tuning plate is fastened to the cavity box, the outer conductor, and is mechanically linked to the front-panel tuning control. This capacity shunts the inner conductor to the outer conductor and varies the electrical length and resonant frequency of the cavity.

2.4 Television Broadcasting

Television transmitters in the U.S. operate in three frequency bands:

- *Low-band VHF.* Channels 2 through 6 (54-72 MHz and 76-88 MHZ).
- *High-band VHF.* Channels 7 through 13 (174-216 MHz).
- *UHF.* Channels 14 through 69 (470-806 MHz). UHF channels 70 through 83 (806-890 MHz) currently are assigned to land mobile radio services. Certain TV translators may continue to operate on these frequencies on a secondary basis.

Because of the wide variety of operating parameters for television stations outside the U. S., this section will focus primarily on TV transmission as it relates to the U.S. (Table 2-1 shows the frequencies used by TV broadcasting). Maximum power output limits are specified by the FCC for each type of service. The maximum effective radiated power for low-band VHF is 100 kW, for high-band VHF it is 316 kW, and for UHF it is 5 MW.

The second major factor that affects the coverage area of a TV station is antenna height, known in the broadcast industry as *height above average terrain* (HAAT). HAAT takes into consideration the effects of the geography in the vicinity of the transmitting tower. The maximum HAAT permitted by the FCC for a low- or high-band VHF station is 1,000 f (305 m) east of the Mississippi River, and 2,000 f (610 m) west of the Mississippi. UHF stations are permitted to operate with a maximum HAAT of 2,000 f (610 m) anywhere in the United States (including Alaska and Hawaii).

Table 2-1 Channel designations for VHF and UHF television stations in the U.S. (*Source:* Fink and Christiansen, *Electronics Engineers' Handbook*, 3rd Ed., McGraw-Hill, New York, 1989.)

Channel designation	Frequency band, MHz	Channel designation	Frequency band, MHz	Channel designation	Frequency band, MHz
2	54–60	30	566–572	57	728–734
3	60–66	31	572–578	58	734–740
4	66–72	32	578–584	59	740–746
5	76–82	33	584–590	60	746–752
6	82–88	34	590–596	61	752–758
7	174–180	35	596–602	62	758–764
8	180–186	36	602–608	63	764–770
9	186–192	37	608–614	64	770–776
10	192–198	38	614–620	65	776–782
11	198–204	39	620–626	66	782–788
12	204–210	40	626–632	67	788–794
13	210–216	41	632–638	68	794–800
14	470–476	42	638–644	69	800–806
15	476–482	43	644–650	70	806–812
16	482–488	44	650–656	71	812–818
17	488–494	45	656–662	72	818–824
18	494–500	46	662–668	73	824–830
19	500–506	47	668–674	74	830–836
20	506–512	48	674–680	75	836–842
21	512–518	49	680–686	76	842–848
22	518–524	50	686–692	77	848–854
23	524–530	51	692–698	78	854–860
24	530–536	52	698–704	79	860–866
25	536–542	53	704–710	80	866–872
26	542–548	54	710–716	81	872–878
27	548–554	55	716–722	82	878–884
28	554–560	56	722–728	83	884–890
29	560–566				

The ratio of visual output power to aural power can vary from one installation to another; however, the aural is typically operated at between 10–20 percent of the visual power. This difference is the result of the reception characteristics of the two signals. Much greater signal strength is required by the consumer's receiver to recover the visual portion of the transmission as opposed to the aural portion. The aural power output is intended to be sufficient for good reception at the fringe of the station's coverage area, but not beyond. It is of no use to a consumer to be able to receive a TV station's audio signal, but not the video.

In addition to the full-power stations discussed previously, two classifications of low power TV stations have been established by the FCC, designed to meet certain community needs. They are:

- *Translators.* Low power systems that rebroadcast the signal of another station on a different channel. Translators are designed to provide "fill-in" coverage for a station that cannot reach a particular community because of the local terrain. Translators operating in the VHF band are limited to 100 W power output (ERP), and UHF translators are limited to 1 kW.
- *Low-Power Television (LPTV).* A service established by the FCC to meet the special needs of particular communities. LPTV stations operating on VHF frequencies are limited to 100 W ERP and UHF stations are limited to 1 kW. LPTV stations originate their own programming and can be assigned by the FCC to any channel, as long as full protection against interference to a full-power station is afforded.

2.4.1   Television Transmission Standards

Television signals transmitted throughout the world have the following similarities:

- All systems use two fields interlaced to create a complete frame.
- All contain luminance, chrominance, sync, and sound components.
- All use amplitude modulation to put picture information onto the visual carrier.
- Modulation polarity, in most cases, is negative (greatest power output from the transmitter occurs during the sync interval, least power output occurs during peak white).
- The sound is transmitted on an aural carrier that is offset on a higher frequency than the visual carrier, using frequency modulation in most cases.

- All systems use a vestigial lower sideband approach.
- All systems derive a luminance and two color difference signals from red, green and blue components.

There the similarities stop and the differences begin. There are three primary color transmission standards in use.

- *NTSC* (National Television Systems Committee). Used in the United States, Canada, Central America, most of South America, and Japan. In addition, NTSC has been accepted for use in various countries or possessions heavily influenced by the United States. The components of the NTSC signal are shown in Figure 2-10.
- *PAL* (Phase Alternation each Line). Used in England, most countries and possessions influenced by the British Commonwealth, many western European countries, and China. Variation exists in PAL systems.
- *SECAM* (SEquential Color with [Avec] Memory). Used in France, countries and possessions influenced by France, the USSR (most Soviet Bloc nations, including East Germany) and other areas influenced by Russia.

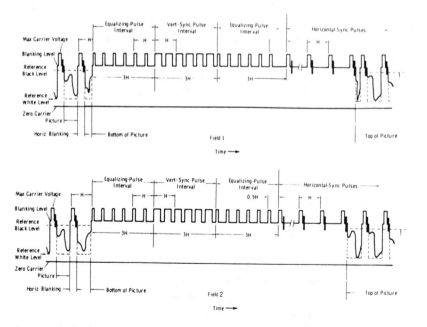

Figure 2-10 The major components of the NTSC television signal. *H* = time from start of line to the start of the next line. *V* = time from the start of one field to the start of the next field.

The three standards are incompatible for the following reasons:

- Channel assignments are made in different frequency spectra in many parts of the world. Some countries have VHF only; some have UHF only; others have both. Assignments with VHF and UHF do not necessarily coincide between countries.
- Channel bandwidths are different. NTSC uses a 6 MHz channel width. Versions of PAL exist with 6 MHz, 7 MHz, and 8 MHz bandwidths. SECAM channels are 8 MHz wide.
- Vision bands are different. NTSC uses 4.2 MHz. PAL uses 4.2 MHz, 5 MHz, and 5.5 MHz. SECAM has 6 MHz video bandwidth.
- The line structure of the signals varies. NTSC uses 525 lines per frame, 30 frames (60 fields) per second. PAL and SECAM use 625 lines per frame, 25 frames (50 fields) per second. As a result, the scanning frequencies also vary.
- The color subcarrier signals are incompatible. NTSC uses 3.579545 MHz. PAL uses 4.43361875 MHz. SECAM utilizes two subcarriers — 4.40625 MHz and 4.250 MHz. The color subcarrier values are derived from the horizontal frequencies in order to interleave color information into the luminance signal without causing undue interference.
- The color encoding system of all three standards differs.
- The offset between visual and aural carriers varies. In NTSC it is 4.5 MHz. In PAL the separation is 5.5 or 6 MHz, depending upon the PAL type. SECAM uses 6.5 MHz separation.
- One form of SECAM uses positive polarity visual modulation (peak white produces greatest power output of transmitter) with amplitude modulation for sound.
- Channels transmitted on UHF frequencies may differ from those on VHF in some forms of PAL and SECAM. Differences include channel bandwidth and video bandwidth.

It is possible to convert from one television standard to another electronically. The most difficult part of the conversion process results from the differing number of scan lines. In general, the signal must be disassembled in the input section of the standards convertor, then placed in a large dynamic memory. Complex computer algorithms compare information on pairs of lines to determine how to create the new lines required (for conversion to PAL or SECAM) or remove lines (for conversion to NTSC). Non-moving objects in the picture present no great difficulties, but motion in the picture can produce objectionable artifacts as the result of the sampling system.

2.4.2   Transmitter Design Considerations

A television transmitter is divided into two basic subsystems: the *visual* section, which accepts the video input, amplitude modulates a radio frequency carrier and amplifies the signal to feed the antenna system; and the *aural* section, which accepts the audio input, frequency modulates a separate RF carrier and amplifies the signal to feed the antenna system. The visual and aural signals are usually combined to feed a single radiating antenna. Different transmitter manufacturers have different philosophies with regard to the design and construction of a transmitter. Some generalizations can, however, be made with respect to basic system design. Transmitters can be divided into categories based on the following criteria:

- Output power
- Final stage design
- Modulation system

Output Power. When the power output of a TV transmitter is discussed, the visual section is the primary consideration. Output power refers to the peak power of the visual stage of the transmitter (peak of sync). The FCC-licensed ERP is equal to the transmitter power output times feedline efficiency times the power gain of the antenna.

A low-band VHF station can achieve its maximum 100 kW power output through a wide range of transmitter and antenna combinations. A 35 kW transmitter coupled with a gain-of-4 antenna would do the trick, as would a 10 kW transmitter feeding an antenna with a gain of 12. Reasonable pairings for a high-band VHF station would range from a transmitter with a power output of 50 kW feeding an antenna with a gain of 8, to a 30 kW transmitter connected to a gain-of-12 antenna. These combinations assume reasonable feedline losses. To reach the exact power level, minor adjustments are made to the power output of the transmitter, usually by a front panel power control.

UHF stations that want to achieve their maximum licensed power output are faced with installing a very high power (and very expensive) transmitter. Typical pairings include a transmitter rated for 220 kW and an antenna with a gain of 25, or a 110 kW transmitter and a gain-of-50 antenna. In the latter case, the antenna could pose a significant problem. UHF antennas with gains in the region of 50 are possible, but not advisable for most installations because of the coverage problems that can result. High gain antennas have a narrow vertical radiation pattern that can reduce a station's coverage in areas near the transmitter site. Whatever way is chosen, getting 5 MW ERP is an expensive proposition. Most UHF stations, therefore, operate considerably below the maximum permitted ERP.

Final Stage Design. The amount of output power required of a transmitter has a fundamental effect on system design. Power levels dictate whether the unit will be of solid-state or vacuum tube design; whether air, water or vapor cooling must be used; the type of power supply required; the sophistication of the high voltage control and supervisory circuitry; and whether common amplification of the visual and aural signals (rather than separate visual and aural amplifiers) is practical.

Tetrodes are generally used for VHF transmitters above 5 kW and for low-power UHF transmitters (below 10 kW). As solid state technology advances, the power levels possible in a reasonable transmitter design steadily increase. As of this writing, all-solid state VHF transmitters of 60 kW have been produced.

In the realm of UHF transmitters, the klystron reigns supreme. Klystrons use an electron bunching technique to generate high power (55 kW from a single tube is not uncommon) at UHF frequencies. They are currently the first choice for high power service. Klystrons, however, are not particularly efficient. A stock klystron with no special circuitry might be only 40 percent efficient. Various schemes have been devised to improve klystron efficiency, the best known of which is beam pulsing. Two types of pulsing are in common use today:

- *Mod-anode pulsing.* A technique designed to reduce power consumption of the device during the color burst and video portion of the signal (and thereby improve overall system efficiency).
- *Annular control electrode* (ACE) pulsing. A technique that accomplishes basically the same thing as mod-anode pulsing by incorporating the pulsing signal into a low-voltage stage of the transmitter, rather than a high-voltage stage (as with mod-anode pulsing).

Experience has shown the newer ACE approach—and other similar designs—to provide greater improvement in operating efficiency than mod-anode pulsing, and better reliability as well.

Several emerging technologies offer additional ways to improve UHF transmitter efficiency. They include:

- The *Klystrode tube*. A device that essentially combines the cathode/grid structure of the tetrode with the drift tube/collector structure of the klystron. (The Klystrode tube is a registered trademark of Varian Associates.) As of this writing, several high power (120 kW or greater) transmitters utilizing Klystrode tube technology have been put into on-air service. Data taken from these systems indicates a significant efficiency improvement over conventional klystron designs.
- The *multi-stage depressed collector* (MSDC) klystron. A device that achieves greater efficiency through a redesign of the collector assembly. A multi-stage collector is used to recover energy from the electron stream inside the klystron and return it to the beam power supply.

Modulation System. A number of approaches may be taken to amplitude modulation of the visual carrier. Current technology systems utilize low level intermediate frequency (IF) modulation. This approach allows superior distortion correction, more accurate vestigial sideband shaping, and significant economic advantages to the transmitter manufacturer.

2.4.3   Elements of the Transmitter

A television transmitter can be divided into four major subsystems:

- The exciter
- Intermediate power amplifier (IPA)
- Power amplifier (PA)
- High voltage power supply

Figure 2-11 shows the audio, video, and RF paths for a typical design. The exciter includes of the following circuits:

- Video input buffer
- Exciter-modulator
- RF processor

Depending on the design of the transmitter, these sections may be separate units or simply incorporated into the exciter itself. A power supply section supplies operating voltages to the various subassemblies of the transmitter.

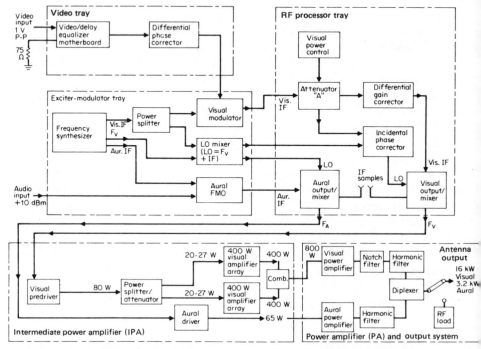

Figure 2-11 Basic block diagram of a television transmitter. The three major subassemblies are the exciter, IPA, and PA. The power supply provides operating voltages to all sections, and high voltage to the PA stage. (*Source:* Benson and Whitaker, *Television and Audio Handbook*, McGraw-Hill, New York, 1990.)

2.4.4 Exciter Subsystem

The input video signal is first fed into a video processing amplifier (*proc amp*) and equalizer. The proc amp performs the following functions:

- Video input amplification
- Group delay equalization/correction
- Video clamping and hum cancellation
- Phase delay compensation
- White clipping

The processed video is converted to amplitude modulation of an IF frequency on a carrier generated by a master oscillator. Major circuits include:

- Frequency synthesizer (or crystal oscillator)
- Visual/aural modulator
- Vestigial sideband filter

The modulated visual IF signal is processed for distortion corrections and subsequent RF transmission in the following stages:

• Power amplifier and driver linearity corrector
• Incidental phase modulation corrector
• RF up-converter
• Channel bandpass filter

Group Delay Equalizer/Corrector. The group delay equalizer is
designed to introduce video delays to offset and correct delays produced in the transmitter PA stage and notch diplexer, and those inherent in typical TV receivers. Each delay equalizer contains up to
five *all-pass* delay circuits, each designed to correct a specific frequency band. Figure 2-12 shows a simplified active all-pass network.

All-pass networks are constant-impedance circuits that allow composite delay curves to be made by cascading sections. The result is
an overall delay curve that complements the transmitter's delay
characteristics. Such circuits can independently correct phase and
amplitude variations with a minimum of interaction. The circuit
shown is a second-order phase rotator in which 360° of phase rotation is achieved. First-order phase rotators may also be used, in
which case either C or L is missing. The maximum phase rotation
for a first-order circuit is 180°.

Video Clamp. The video clamp is required to hold the back porch
level constant, regardless of changes in the average picture level. A
sample of the input signal is sent to the clamp circuit, where a
clamp-pulse generator produces an output that is coincident with the

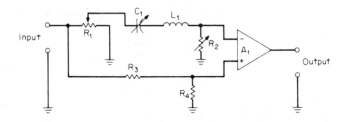

Figure 2-12 Simplified diagram of an active all-pass network, used in a TV
transmitter for group delay correction. In a typical system, five such
circuits would be cascaded to provide the necessary range of correction.
(*Source:* Harris Corp. in K. Blair Benson, *Television Engineering
Handbook,* McGraw-Hill, New York, 1986.)

back porch of the video signal. After removal of the color subcarrier, a separate signal from the output of the video processing system is applied to a tip-of-sync detector. These two signals are processed to produce a control voltage that holds the back porch level constant. In the event of a loss of video, the circuit will automatically set to the blanking level, preventing overdrive of subsequent amplifiers.

Differential Phase Corrector. Differential phase correction is performed on a *precorrection basis* to compensate for the distortion that will be produced in the IPA and PA stages of the transmitter. The corrector functions by generating signals of opposite phase delay which, when added to the input video signal, will cancel subsequent unwanted differential-phase distortion. Figure 2-13 shows a simplified diagram of a differential phase corrector. In a typical implementation, seven such diode gating circuits would be used. This type of configuration permits up to a 14° correction. An examination of this figure will help to understand the concept involved. The 10 kΩ differential phase adjustment potentiometer produces a bias of up to 6.2 V on zener diodes CR1 and CR2. Depending on the bias value, the zener diodes will conduct over a portion of the video signal, with a correction signal being added to the output through the 7 pF capacitor. An inverted input signal, produced by an amplifier not shown on the diagram, is added to the output via diode CR3.

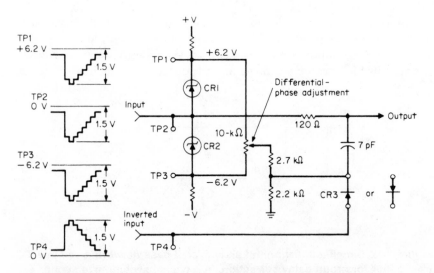

Figure 2-13 Simplified diagram of a differential phase corrector and a set of representative input waveforms. The inverted input at TP4 requires an inverting amplifier, not shown. (*Source:* RCA Corp. in K. Blair Benson, *Television Engineering Handbook,* McGraw-Hill, New York, 1986.)

White Clip. A white-clip circuit is used to prevent negative-going video modulation from dropping below 12.5 percent of reference white. If the video is below 12.5 percent, over-modulation and high incidental (intercarrier) phase modulation (ICPM) will occur. Audio buzz may also be encountered in home receivers. The white clip circuit acts on the active video signal and keeps it from reaching reference white. In so doing, it prevents the transmitter from over-modulating and degrading the audio.

2.4.5 Frequency Synthesizer

The carrier signal for the visual and aural outputs of a TV transmitter begins with a master oscillator. This oscillator can be either crystal-based or generated by a frequency synthesizer. The latter is by far the most common today.

A frequency synthesizer is built around a phased-locked-loop (PLL) circuit that accurately controls the operating frequency of a free-running L/C oscillator. The PLL uses a highly stable crystal oscillator (mounted in a temperature-controlled oven) or a temperature compensated crystal oscillator (TCXO) as a reference signal source. Depending upon the design, a single frequency synthesizer circuit may be used to cover the entire VHF or UHF operating band. Any single operating frequency can be selected by setting DIP switches on the synthesizer circuit board.

The heart of a frequency synthesizer is the PLL, shown in Figure 2-14 in simplified form. In this example, a 5 kH reference clock is fed to the phase detector, where it is compared with a sample of the

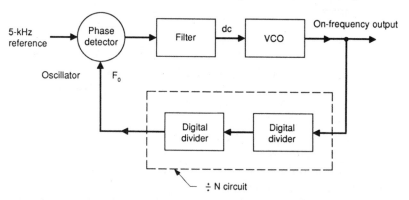

Figure 2-14 Block diagram of a phase locked loop (PLL) frequency synthesizer. This type of circuit is commonly used to generate the operating frequencies in a TV transmitter exciter. (*Source:* RCA Corp. in K. Blair Benson, *Television Engineering Handbook,* McGraw-Hill, New York, 1986.)

final IF carrier, which has been divided down to 5 kHz. If the two signals differ in phase or frequency, an error voltage proportional to the difference is generated and applied to the voltage-controlled oscillator (VCO). This error voltage forces the VCO to move back to the desired operating frequency. The output of the VCO feeds a string of digital counters/dividers that determine the operating frequency of the system. By selecting different "divide by" configurations, the operating point of the VCO can be made to change.

The output signal of the frequency synthesizer at visual IF is applied to the visual modulator. A second visual output plus the visual carrier is used for incidental phase correction and up-conversion. A similar signal is also sent to the aural up-converter. Although the aural system of a television transmitter is essentially independent from the visual (in most designs), the aural carrier must be accurately locked to the visual.

The need to keep the visual and aural carriers locked together is a function of the basic television system design. In the NTSC system, the color subcarrier is an odd multiple of half the line scanning rate, making the visibility of the subcarrier minimal on monochrome receivers. Developers of the NTSC system determined that optimum performance would be gained if the beat between the subcarrier and the aural carrier was also an odd multiple of half the line scanning rate, rendering it minimally visible. To meet this requirement, aural carrier separation must be a multiple of the horizontal scanning rate. Therefore, although the visual and aural sections of a transmitter can be considered as separate systems, the carrier frequencies of the two systems must be linked together and maintained within tight tolerances.

2.4.6   Visual/Aural Modulator

The visual modulator receives a precorrected video signal from the processing system and uses it to modulate a carrier from the IF synthesizer. Mixer designs vary from one manufacturer to another, but all perform essentially the same function. Figure 2-15 shows a modulator built around a balanced mixer. The resulting amplitude modulated signal is produced with double-sidebands. A downstream filter in the modulator removes most of the lower sideband, resulting in a vestigial sideband signal.

The aural modulator functions essentially in the same fashion as an FM radio broadcast exciter. The peak deviation of the FM aural carrier is 25 kHz (FM broadcast utilizes 75 kHz deviation). Audio input is pre-emphasized to match the standard 75 µs pre-emphasis

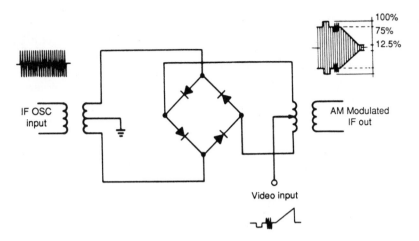

Figure 2-15 Simplified schematic diagram of a balanced mixer. This design is typically used in the visual modulator section of a TV exciter. (*Source:* Benson and Whitaker, *Television and Audio Handbook*, McGraw-Hill, New York, 1990.)

curve specified by the FCC. In most 625 line countries, 50 µs pre-emphasis is used, with 50 kHz deviation. The aural modulator signal frequency-modulates an IF carrier output by the master frequency synthesizer. This signal is then routed to the aural up-converter.

Vestigial Sideband Filter. The modulated visual IF signal is band-shaped in a vestigial sideband filter, typically a *surface-acoustic-wave* (SAW) filter. Envelope delay correction is not required for the SAW filter because of the uniform delay characteristics of such devices. Envelope-delay compensation may, however, be needed for other parts of the transmitter.

The SAW filter provides many benefits to transmitter designers and operators. A SAW filter requires no adjustments and is stable with respect to temperature and time. SAW filters are made of a substrate about 2 mm thick and generally use a piezoelectric crystal such as lithium tantalate (other crystalline structures are also used successfully). The dimensions of the filter elements are determined by the desired frequency response characteristics of the crystal.

SAW filters have exceptionally steep skirts at band edges and can provide up to 60 dB attenuation to out-of-band frequencies. Insertion loss for a SAW filter is about 25 dB, making it necessary to use low noise amplifiers to maintain a satisfactory signal-to-noise ratio for the overall system.

Linearity Corrector. The linearity corrector predistorts the modulated IF signal to compensate for transmitter differential-gain distortion. This is accomplished by compressing or stretching the nonlinear portions of the transfer characteristic of an IF-signal amplifier. The approach is similar to the method used for gamma correction in a video amplifier.

ICPM Corrector. The *incidental phase modulation* (ICPM) corrector cancels ICPM distortion generated within the visual RF path. A sample of the modulated visual carrier is detected to develop a video signal that contains all delay and differential phase pre-corrections. This signal is filtered and used to phase-modulate the local oscillator feed to the visual up-converter. This modulation is equal to and opposite in phase from the undesired ICPM. Once correction has been obtained, envelope and synchronous demodulators in home receivers should produce essentially the same output waveforms. This attribute is important for the growing numbers stereo receivers being purchased by consumers.

Up-Converter. The modulated IF signal is mixed with an output from the local oscillator that is the sum of the IF carrier and the operating carrier frequency. Sum and difference frequency signals are produced in the mixing process and a final carrier difference signal is extracted through the use of a bandpass filter. The visual output of the mixer (up-converter) at the final carrier frequency is sent to the IPA. A similar process occurs in the aural signal path.

2.4.7   Intermediate Power Amplifier

The function of the IPA is to develop the power output necessary to drive the aural and visual systems' power amplifier stages. A low-band 16- to 20-kW transmitter typically requires about 800 W RF drive, and a high-band 35- to 50-kW transmitter needs about 1,600 W. A UHF transmitter utilizing a high gain klystron final tube requires about 20 W drive, while a UHF transmitter utilizing a Klystrode tube needs about 80 W. Because the aural portion of a television transmitter operates at only 10 to 20 percent of the visual power output, the RF drive requirements are proportionately lower.

Virtually all transmitters manufactured today utilize solid state devices in the IPA. Transistors are preferred because of their inherent stability, reliability, and ability to cover a broad band of frequencies without retuning. Present solid state technology, however, cannot provide the power levels needed by most transmitters in a single device. To achieve the needed RF energy, devices are combined using a variety of schemes.

A typical "building block" for a solid state IPA provides a maximum power output of approximately 200 W. In order to meet the requirements of a 20 kW low-band VHF transmitter, a minimum of four such units would have to be combined. In actual practice, some amount of *headroom* is always designed into the system to compensate for component aging, imperfect tuning in the PA stage, and device failure. Most solid state IPA circuits are configured so that in the event of a failure in one module, the remaining modules will continue to operate. If sufficient headroom has been provided in the design, the transmitter will continue to operate without change. The defective subassembly can then be repaired and returned to service at a convenient time.

Because the output of the RF up-converter is about 10 W, an intermediate amplifier is generally used to produce the required drive for the parallel amplifiers. The individual power blocks are fed by a splitter that feeds equal RF drive to each unit. The output of each RF power block is fed to a hybrid combiner that provides isolation between the individual units. The combiner feeds a bandpass filter that allows only the modulated carrier and its sidebands to pass.

The inherent design of a solid state RF amplifier permits operation over a wide range of frequencies. Most drivers are broadband and require no tuning. Certain frequency-determined components are added at the factory (depending upon the design), however, from the end-user standpoint, solid state drivers require virtually no attention. IPA systems are available that cover the entire low- or high-band VHF channels without tuning.

Advances continue to be made in solid state RF devices. New developments promise to substantially extend the reach of semiconductors into medium-power RF operation. Coupled with better devices are better circuit designs, including parallel components and new push-pull configurations. Another significant factor in achieving high power from a solid state device is efficient removal of heat from the component itself.

2.4.8 Power Amplifier

The power amplifier (PA) raises the output energy of the transmitter to the required RF operating level. As noted previously, solid state devices are increasing in use through parallel configurations in high power transmitters. Still, however, the majority of television transmitters in use today utilize vacuum tubes. The workhorse of VHF television is the tetrode, which provides high output power, good efficiency, and good reliability. In UHF service, the klystron is the standard output device for transmitters above 10 kW.

Tetrodes in television service are operated in the class B mode to obtain reasonable efficiency while maintaining a linear transfer characteristic. Class B amplifiers, when operated in tuned circuits, provide linear performance because of the flywheel effect of the resonance circuit. This allows a single tube to be used instead of two in push-pull fashion. The bias point of the linear amplifier must be chosen so that the transfer characteristic at low modulation levels matches that at higher modulation levels. Even so, some nonlinearity is generated in the final stage, requiring differential gain correction. The plate (anode) circuit of a tetrode PA is usually built around a coaxial resonant cavity, which provides a stable and reliable tank.

UHF transmitters using a klystron in the final output stage must operate at class A, the most linear but also most the inefficient operating mode for a vacuum tube. The basic efficiency of a non-pulsed klystron is approximately 40 percent. Pulsing, which provides full available beam current only when it is needed (during peak of sync), can improve device efficiency by as much as 25 percent, depending on the type of pulsing used. (As discussed in Section 2.4.2, recent developments in UHF device technology promise to radically change the efficiency penalty now faced by UHF broadcasters.)

Two types of klystrons are presently in service: integral cavity and external cavity devices. The basic theory of operation is identical for each tube, however, the mechanical approach is radically different. In the integral cavity klystron, the cavities are built into the klystron to form a single unit. In the external cavity klystron, the cavities are outside the vacuum envelope and bolted around the tube when the klystron is installed in the transmitter. A number of factors come into play in a discussion of the relative merits of integral vs. external cavity designs. Primary considerations include operating efficiency, purchase price, and life expectancy.

The PA stage includes a number of sensors that provide input to supervisory and control circuits. Because of the power levels present in the PA stage, sophisticated fault-detection circuits are required to prevent damage to components in the event of a problem either external to, or inside the transmitter. An RF sample, obtained from a directional coupler installed at the output of the transmitter, is used to provide automatic power-level control.

The transmitter system discussed so far assumes separate visual and aural PA stages. This configuration is normally used for high power transmitters. Low power designs often use a combined mode in which the aural and visual signals are added prior to the PA. This approach offers a simplified system, but at the cost of additional pre-correction of the input video signal. PA stages often are configured so that the circuitry of the visual and aural amplifiers are identical. While this represents a good deal of "overkill" insofar as the aural PA is concerned, it provides backup protection in the event of a visual PA failure. The aural PA can then be reconfigured to amplify both the aural and the visual signals, at reduced power.

The aural output stage of a television transmitter is similar in basic design to an FM broadcast transmitter. Tetrode output devices generally operate at class C, providing good efficiency. Klystron-based aural PAs are used in UHF transmitters.

2.4.9   Coupling/Filtering System

The output of the aural and visual power amplifiers must be combined and filtered to provide a signal that is electrically ready to be applied to the antenna system. The primary elements of the coupling and filtering system of a TV transmitter are:

- Color notch filter
- Aural and visual harmonic filters
- Diplexer

In a low power transmitter (below 5 kW), this hardware may be included within the transmitter cabinet itself. Normally, however, it is located externally to the transmitter.

Color Notch Filter. The color notch filter is used to attenuate the color subcarrier lower sideband to the -42 dB requirements of the FCC. The color notch filter is placed across the transmitter output feedline. The filter consists of a coax or waveguide stub tuned to 3.58 MHz below the picture carrier. The Q of the filter is high enough so that energy in the vestigial sideband is not materially affected, while still providing high attenuation at 3.58 MHz.

Harmonic Filters. To ensure compliance with FCC requirements, harmonic filters are used to attenuate out-of-band radiation of the aural and visual signals. Filter designs vary depending upon the manufacturer; however, most are of coaxial construction utilizing components housed within a prepackaged assembly. Stub filters are also used, typically adjusted to provide maximum attenuation at the second harmonic of the operating frequency of the visual and the aural carriers.

Diplexer/Combiner. The filtered visual and aural outputs are fed to a diplexer where the two signals are combined to feed the antenna

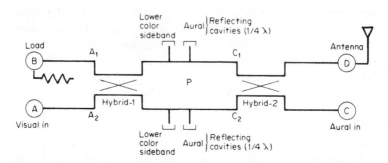

Figure 2-16 Functional diagram of a notch diplexer, used to combine the aural and visual outputs of a television transmitter for application to the antenna. (*Source:* Micro Communications, Inc. In K. Blair Benson, *Television Engineering Handbook,* **McGraw-Hill, New York, 1986.)**

(see Figure 2-16). For installations that require dual-antenna feedlines, a hybrid combiner with quadrature-phased outputs is used. Depending upon the design and operating power, the color notch filter, aural and visual harmonic filters, and diplexer may be combined into a single mechanical unit.

A hybrid combiner serves as the building block of the notch diplexer, which combines the aural and visual RF signals to feed a single line antenna system and provide a constant impedance load to each section of the transmitter. The notch diplexer consists of two hybrid combiners and two sets of reject cavities. The system is configured so that all of the energy from the visual transmitter passes to the antenna (port D), and all of the energy from the aural transmitter passes to the antenna. The phase relationships are arranged so that the input signals cancel at the resistive load (port B). Because of the paths taken by the aural and visual signals through the notch diplexer, the amplitude and phase characteristics of each input do not change from the input ports (port A for the visual and port C for the aural) and the antenna (port D), thus preserving signal purity.

Bibliography

1. Benson, K., *Television Engineering Handbook,* McGraw-Hill, New York, 1986.
2. Benson, K., and J. Whitaker, *Television and Audio Handbook for Technicians and Engineers,* McGraw-Hill, New York, 1989.
3. Fink, D., and D. Christiansen, *Electronics Engineer's Handbook,* Third Edition, McGraw-Hill, New York, 1989.

3

Non-Broadcast Applications

3.1 Introduction

Radio and TV broadcasting are the most obvious applications of RF technology. In total numbers of installations, however, non-broadcast uses for RF far out-distance radio and TV stations. Applications range from microwave communications to induction heating. Power levels range from a few tens of Watts to a million Watts or more. The areas of non-broadcast RF technology covered in this chapter include:

- Satellite transmission
- Radar
- Electronic navigation
- Induction heating

3.2 Satellite Transmission

Commercial satellite communication began on July 10, 1962, when television pictures were first beamed across the Atlantic Ocean through the Telstar 1 satellite. Three years later, the INTELSAT system of *geostationary* relay satellites saw its initial craft, *Early Bird 1,* launched into a rapidly growing communications industry. In the same year, the USSR inaugurated the Molnya series of satellites travelling in an elliptical orbit to better meet the needs of that nation. The Molnya satellites were placed in an orbit inclined about

64°, relative to the equator, with an orbital period half that of the Earth.

All commercial satellites in use today operate in a geostationary orbit. A geostationary satellite is one that maintains a fixed position in space relative to Earth because of its altitude, roughly 22,300 miles above the Earth. Two primary frequency bands are used, the *C-band* (4-6 GHz) and the *Ku-band* (11-14 GHz). Any satellite relay system involves three basic sections:

* An *uplink* transmitting station, which beams signals toward the satellite in its geostationary orbit.
* The *satellite* (the space segment of the system), which receives, amplifies and retransmits the signals back to Earth.
* The *downlink* receiving station, which completes the relay path.

Because of the frequencies involved, satellite communication is designated as a microwave radio service. As such, certain requirements are placed upon the system. Like terrestrial microwave, the path between transmitter and receiver must be line-of-sight. Meteorological conditions, such as rain and fog, result in a detrimental attenuation of the signal. Arrangements must be made to shield satellite receive antennas from terrestrial interference. Because received signal strength is based upon the inverse square law, highly directional transmit and receive parabolic antennas are used, which in turn require a high degree of aiming accuracy. To counteract the effects of galactic and thermal noise sources on low level signals, amplifiers are designed for exceptionally low noise characteristics. Figure 3-1 shows the primary elements of a satellite relay system.

3.2.1 Satellite Communications

The first satellites launched for INTELSAT and other users contained only one or two *transponders*, radio relay units. Pressure for increased satellite link services has driven engineers to develop more economical systems with multiple transponder designs. Generally, C-band satellites placed in orbit now have 12 or 24 transponders, each with 36 MHz bandwidths. Ku-band systems often use fewer transponders with wider bandwidths.

Users of satellite communication links are assigned to transponders generally on a lease basis, although it may be possible to purchase a transponder. Assignments usually leave one or more spare transponders aboard each craft, allowing for standby capabilities in the event a transponder should fail.

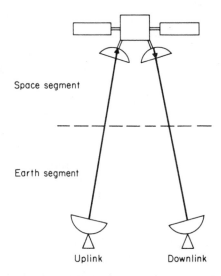

Space segment

Earth segment

Uplink   Downlink

Figure 3-1 A satellite communications link consists of an uplink, the satellite (as the space segment), and a downlink. (*Source:* Benson and Whitaker, *Television and Audio Handbook,* McGraw-Hill, New York, 1990.)

By assigning transponders to users, the satellite operator simplifies the design of uplink and downlink facilities. An Earth station controller can be programmed according to the transponder of interest. For example, a corporate video facility may need to access four or five different transponders from one satellite. To do so, the operator needs only to enter the transponder number (or carrier frequency) of interest. The receiver handles retuning and automatic switching of signals from a dual polarity feedhorn on the antenna.

Each transponder has a fixed center frequency and a specific signal polarization. For example, according to current frequency allocation plans, all odd numbered transponders use horizontal polarization, while the even numbered ones use vertical polarization. Excessive deviation from the center carrier frequency by one signal does not cause interference to another transponder because of the isolation provided by cross-polarization. This concept is extended to satellites in adjacent parking spaces in *geosynchronous* orbit. Center frequencies for transponders on adjacent satellites are offset in frequency from those on the first craft. In addition, an angular offset of polarization is employed. The even and odd transponder assignments are still offset by 90° from one another. As spacing decreases between satellites, the polarization offset must increase to reduce the potential for interference.

3.2.2   Satellite Uplink

The ground-based transmitting equipment for a satellite system consists of three sections: baseband, intermediate frequency (IF) and radio frequency (RF).

The baseband section interfaces various incoming signals with the transmission format of the satellite being used. Signals provided to the baseband section may already be in a modulated form with modulation characteristics (digital, analog, or some other format) determined by the terrestrial media that brings the signals to the uplink site. Depending upon the nature of the incoming signal (voice, data, or video), some degree of processing will be applied. In many cases, multiple incoming signals will be combined into a single composite uplink signal through multiplexing.

When the incoming signals are in the correct format for transmission to the satellite, they are applied to an FM modulator, which converts the composite signal upward to a 70 MHz intermediate frequency. The use of an IF section has several advantages:

- A direct conversion between baseband and the output frequency presents difficulties in maintaining the frequency stability of the output signal.
- Any mixing or modulation step has the potential of introducing unwanted byproducts. Filtering at the IF may be used to remove spurious signals resulting from the mixing process.
- Many terrestrial microwave systems include a 70 MHz IF section. If a signal is brought into the uplink site by a terrestrial microwave, it becomes a simple matter to connect the signal directly into the IF section of the uplink system.

From the 70 MHz IF, the signal is converted upward again, this time to the output frequency of 6 GHz (for C-band) or 14 GHz (for Ku-band) before application to a high power amplifier (HPA). Conventional Earth station transmitters operate over a wide power range, from a few tens of Watts to 12 kW. Transmitters designed for deep space research may operate at up to 400 kW.

Several amplifying devices are used in HPA designs, depending upon the power output and frequency requirements. For the highest power level in C- or Ku-band, klystrons are employed. Devices are available with pulsed outputs ranging from 500 W to 5 kW, and a bandwidth capability of 40 MHz. This means that a separate klystron is required for each 40 MHz wide signal to be beamed upward to a transponder.

The *traveling wave tube* (TWT) is another type of vacuum power device used for HPA transmitters. While similar to some areas of klystron operation, the TWT is capable of amplifying a band of signals at least ten times wider than the klystron. Thus, one TWT system can be used to amplify the signals sent to several transponders on the satellite. With output powers from 100 W to 2.5 kW, the bandwidth capability of the TWT offsets its much higher cost than the klystron in some applications.

Solid-state amplifiers based on MOSFET technology can be used for both C and Ku-band uplink HPA systems. The power capabilities of solid-state units are limited: 5-50 W for C-band and 1-6 W for Ku-band. Such systems, however, offer wideband performance and good reliability.

Uplink Antennas. The output of the HPA, when applied to a parabolic reflector antenna, experiences a high degree of gain when referenced to an ideal isotropic antenna (dBi). For example, large reflector antennas approximately 10 m in diameter offer gains as high as 55 dB, increasing the output of a 3 kW klystron or TWT amplifier to an effective radiated power of 57-86 dBW. Smaller reflector sizes (6-8 m) may also be used, with the observation of certain restrictions in regard to interference with other satellites and other services. Not surprisingly, smaller antennas provide lower gain. For a 30 m reflector, such as those used for international satellite communications, approximately 58 dB gain may be achieved. Several variations of parabolic antenna designs are used for satellite communications services, including the following (see Figure 3-2):

- *Prime focus, single parabolic reflector.* Places the source of the signal to be transmitted in front of the reflector precisely at the focal point of the parabola. Large antennas of this type commonly employ a feed horn supported with a tripod of struts. Because the struts (waveguide to the feedhorn and the horn assembly itself) are located directly within the transmitted beam, every effort is made to design these components with as little bulk as possible, yet physically strong enough to withstand adverse weather conditions.
- *Offset reflector.* Removes the feed horn and its support from the radiated beam. Although the reflector maintains the shape of a section of the parabola, the closed end of the curve is not included. The feed horn, while still located at the focal point of the curve, points at an angle from the vertex of the parabola shape.
- *Double reflector.* The primary reflector is parabolic in shape while the subreflector surface, mounted in front of the focal point of the

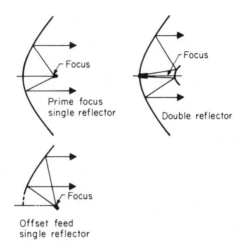

Figure 3-2 Satellite transmitting/receiving antennas are of three general types: prime focus, single reflector; offset feed, single reflector; and double reflector. (*Source:* Benson and Whitaker, *Television and Audio Handbook*, McGraw-Hill, New York, 1990.)

parabola, is hyperbolic in shape. One focus of the hyperbolic reflector is located at the parabolic focal point, while the second focal point of the subreflector defines the position for the feed horn signal source. Signals reflected from the hyperbolic subreflector are spread across the parabolic prime reflector, which then directs them as a parallel beam toward the satellite. This two-reflector antenna provides several advantages over a single-reflector type:

— The overall front-to-back dimension of the two-reflector system is shorter, which simplifies mounting and decreases wind-loading.
— Placement of the subreflector closer to the main reflector generates less spill-over signal because energy is not directed as closely to the edge of the main reflector.
— Accuracy of the reflector surfaces is not as stringent as with a single-reflector type of structure.

The antenna used for transmission of signals to the satellite can also be used to receive signals from the satellite. The major change needed to provide this capability is the addition of directional switching or coupling to prevent HPA transmitter energy from entering the receiver system. Switching devices, or *circulators*, use waveguide characteristics to create a signal path linking the transmitter signal to the antenna feedhorn, while simultaneously providing a received signal path from the feed horn to the receiver input.

Signal Formats. The signal transmitted from the uplink site (or from the satellite, for that matter) is in the form of frequency modulation. Limitations are placed on uplinked signals to avoid interference problems resulting from excessive bandwidth. For example, a satellite relay channel for television use typically contains only a single video signal and its associated audio. Audio is carried on one or more subcarriers that are stacked onto the video signal. To develop the composite signal, each audio channel is first modulated onto its subcarrier frequency. Then, each of the subcarriers and the main channel of video are applied as modulation to the uplink carrier. The maximum level of each component is controlled to avoid overmodulation.

In the case of telephone relay circuits, the same subcarrier concept is used. A number of individual voice circuits are combined into groups, which are then multiplexed to subcarriers through various digital means. The result is that thousands of telephone conversations may occur simultaneously through a single satellite.

3.2.3 Satellite Link

Like other relay stations, the communications spacecraft contains antennas for receiving and retransmission. From the antenna, signals pass through a low noise amplifier before frequency conversion to the transmit band. A high power amplifier feeds the received signal to a directional antenna, which beams the information to a predetermined area of the Earth to be served by the satellite (see Figure 3-3). Power to operate the electronics hardware is generated by solar

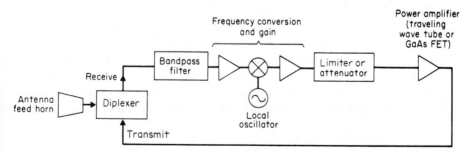

Figure 3-3 Block diagram of a satellite transponder channel. (*Source:* Fink and Christiansen, *Electronics Engineers' Handbook, 3rd Ed.,* McGraw-Hill, New York, 1989.)

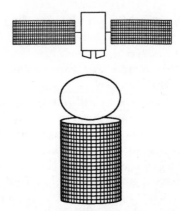

Figure 3-4 The two most common types of solar cell arrays used for communications satellites. (*Source:* Benson and Whitaker, *Television and Audio Handbook,* McGraw-Hill, New York, 1990.)

cells. Inside the satellite, storage batteries kept recharged by the solar cell arrays carry the electronic load when the satellite is eclipsed by the Earth. Figure 3-4 shows the two most common solar cell configurations. Power to the electronics on the craft requires protective regulation to maintain consistent signal levels. Most of the equipment operates at low voltages, but the final stage of each transponder chain ends in a high power amplifier. The HPA of C-band satellite channels may include a traveling wave tube or a solid-state power amplifier (SSPA). Ku-band systems rely primarily on TWT devices at present. Klystrons and TWTs require multiple voltages levels. The filaments operate at low voltages, but beam focus and electron collection electrodes require voltages in the hundreds and thousands of volts. To develop such a range of voltages, the satellite power supply includes voltage converters.

From these potentials, the klystron or TWT will produce output powers in the range of 8.5 W to 20 W. Most systems are operated at the lower end of the range to increase reliability and life expectancy. In general, the lifetime of the spacecraft is assumed to be seven years. A guidance system is included to stabilize the attitude of the craft as it rotates around the Earth. Small rocket engines are provided for maintaining an exact position in the assigned geostationary arc (see Figure 3-5). This work is known as *station-keeping.*

Satellite Antennas. The antenna system for a communications satellite is really several antennas combined into a single assembly. One is for receiving signals from Earth. Another, obviously, is for

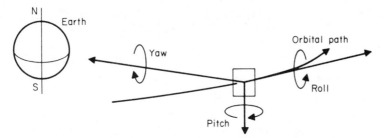

Figure 3-5 Attitude of the spacecraft is determined by pitch, roll, and yaw, rotations around three references axes. (*Source:* Benson and Whitaker, *Television and Audio Handbook,* McGraw-Hill, New York, 1990.)

transmitting those signals back to Earth. The transmitting antenna may be made of more than one section to handle the needs of multiple signal beams. Finally, a receive-transmit beacon antenna provides communication with the ground-based satellite control station.

At the receiving end of the transponder, signals coming from the antenna are split into separate bands through a channelizing network, allowing each input signal to be directed to its own receiver, processing amplifier, and HPA. At the output, a combiner brings all channels together again into one signal to be fed to the transmitting antenna.

The approach to designing the complex antenna system for a relay satellite depends a good deal on horizontal and vertical polarization of the signals as a means to keep incoming and outgoing information separated. Multilayer, dichroic reflectors that are sensitive to the polarizations can be used for such purposes. Also, multiple feed horns may be needed to develop one or more beams back to Earth. Antennas for different requirements may combine several antenna designs, but nearly all are based upon the parabolic reflector. The parabolic design offers a number of unique properties. First, rays received by such a structure that are parallel to the feed axis are reflected and converged at the focus. Second, rays emitted from the focal point are reflected and emerge parallel to the feed axis. Special cases may involve some use of spherical and elliptical reflector shapes, but the parabolic is of most importance.

3.2.4   Satellite Downlink

Satellite receiving stations, like uplink equipment, perform the function of interfacing ground-based equipment to satellite transponders. Earth stations consist of a receiving antenna, *low noise amplifier,* 4

GHz (C-band) or 11 GHz (Ku-band) tuner, 70 MHz IF section, and baseband output stage.

Downlink Antennas. Antenna type and size for any application is determined by the mode of transmission, band of operation, location of the receiving station, typical weather in the receiving station locale, and the required quality of the output signal. Digital transmissions allow a smaller main reflector to be used, because the decoding equipment is usually designed to provide error correction. The data stream periodically includes information to check the accuracy of the data, and if errors are found, to make corrections. Sophisticated error concealment techniques make it possible to hide errors that are too large for correction. Greater emphasis is placed on error correction for applications involving financial transactions or life-critical data, such as might be involved with a manned space flight. For entertainment programming, such as TV broadcasts and audio, absolute correction is less critical and gives way primarily to concealment techniques.

Receiving antennas for commercial applications, such as radio/TV networks, cable TV networks, and special services or teleconferencing centers, generally fall into the 7–10 m range for C-band operation. Ku-band units can be smaller. Antennas for consumer and business use may be even more compact, depending upon the type of signal being received and the quality of the signal to be provided by the downlink. The nature of the application also helps to determine if the antenna will be strictly parabolic, or if one of the spherical types, generally designed for consumer use, will be sufficient.

In general, the gain and directivity of a large reflector is greater than a small reflector. The size of the reflector required depends upon the level of signal that can be reliably received at a specific location under the worst possible conditions. Gain must be adequate to bring the RF signal from the satellite to a level that is acceptable to the electronics equipment. The output signal must maintain a signal-to-noise ratio that is sufficiently high so the receiver electronics can recover the desired signal without significant degradation from noise.

It is instructive to consider the power budget of the downlink, that is, a calculation of positive and negative factors determining signal level. Figure 3-6 shows an analysis of both the uplink and downlink functions, and typical values of gain or loss. From this figure the need for receiving equipment with exceptional low noise performance becomes more obvious. One of the most critical parts of the receiver

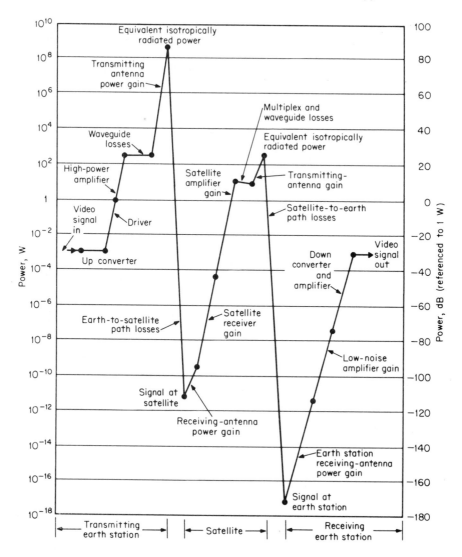

Figure 3-6 Typical values of gain and loss in the transmission of a TV signal via satellite. (*Source:* B. I. Edelson, "Global Satellite Communications," *Scientific American,* February 1977. Copyright 1977 by Scientific American, Inc. All rights reserved. In Fink and Christiansen, *Electronic Engineers' Handbook, 3rd Ed.,* McGraw-Hill, New York, 1989.)

is the *low noise amplifier* (LNA) or *low noise conversion unit* (LNC), which is the first component following the antenna to process the signal. Such devices are rated by their *noise temperature,* usually a number around 211°K. The cost of a LNA or LNC increases significantly as the temperature figure goes down.

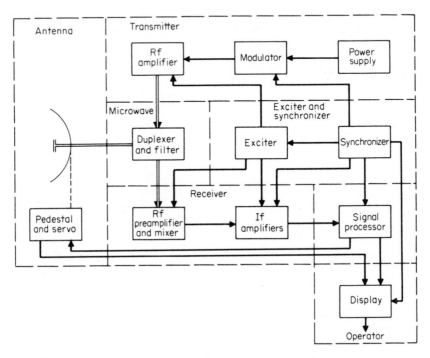

Figure 3-7 Simplified block diagram of a pulsed radar system. (*Source:* Fink and Christiansen, *Electronics Engineers' Handbook, 3rd Ed.,* McGraw-Hill, New York, 1989.)

3.3 Radar[1]

The word radar is an acronym for *radio detection and ranging.* The name accurately spells out the basic function of a radar system. Measurement of target angles is an additional function of most radar equipment. Doppler velocity may also be measured as an important parameter. A block diagram of a typical pulsed radar system is shown in Figure 3-7. Any system can be divided into six basic subsections:

• Exciter and synchronizer. Controls the sequence of transmission and reception.

---

1 Portions of Section 3.3 (Radar) and Section 3.4 (Electronic Navigation) were adapted from *The Electronics Engineers' Handbook,* Third Edition, Section 25, "Radar, Navigation and Underwater Sound Systems," Donald G. Fink and Donald Christiansen editors.

**Table 3-1 Typical radar applications. (*Source:* Fink and Christiansen,
Electronics Engineers' Handbook, 3rd Ed., McGraw-Hill, New York, 1989.)**

Air surveillance	Long-range early warning, ground-controlled intercept, acquisition for weapon system, height finding and three-dimensional radar, airport and air-route surveillance
Space and missile surveillance	Ballistic missile warning, missile acquisition, satellite surveillance
Surface-search and battlefield surveillance	Sea search and navigation, ground mapping, mortar and artillery location, airport taxiway control
Weather radar	Observation and prediction, weather avoidance (aircraft), cloud-visibility indicators
Tracking and guidance	Antiaircraft fire control, surface fire control, missile guidance, range instrumentation, satellite instrumentation, precision approach and landing
Astronomy and geodesy	Planetary observation, earth survey, ionospheric sounding

- Transmitter. Generates a high power RF pulse of specified frequency and shape.
- Microwave network. Couples the transmitter and receiver sections to the antenna.
- Antenna system. Consists of a radiating/receiving structure mounted on a mechanically-steered servo-driven pedestal. A *stationary array,* which uses electrical steering of the antenna system, may be used in place of the mechanical system shown in the figure.
- Receiver. Selects and amplifies the return pulse picked up by the antenna.
- Signal processor and display. Integrates the detected echo pulse, synchronizer data, and antenna pointing data for presentation to an operator.

Radar technology is used for countless applications. Table 3-1 lists some of the more common uses.

Table 3-2 Radar frequency bands (IEEE standard 521-1976). (*Source:* Fink and Christiansen, *Electronics Engineers' Handbook, 3rd Ed.,* McGraw-Hill, New York, 1989.)

Name	Frequency range	Radiolocation bands based on ITU assignments in region II
VHF	30–300 MHz	137–144 MHz
UHF	300–1,000 MHz	216–225 MHz
P band†	230–1,000 MHz	420–450 MHz 890–940* MHz
L band	1,000–2,000 MHz	1,215–1,400 MHz
S band	2,000–4,000 MHz	2,300–2,550 MHz 2,700–3,700 MHz
C band	4,000–8,000 MHz	5,255–5,925 MHz
X band	8,000–12,5000 MHz	8,500–10,700 MHz
K_u band	12.5–18 GHz	13.4–14.4 GHz 15.7–17.7 GHz
K band	18–26.5 GHz	23–24.25 MHz
K_a band	26.5–40 GHz	33.4–36 MHz
Millimeter	> 40 GHz	

* Sometimes included in L band.
† Seldom used nomenclature.

3.3.1   Radar Parameters

Because radar systems have many diverse applications, the parameters of frequency, power, and transmission format also vary widely. There are no fundamental bounds on the operating frequencies of radar. In fact, any system that locates objects by detecting echoes scattered from a target that has been illuminated with electromagnetic energy can be considered radar. While the principles of operation are similar regardless of the frequency, the functions and circuit parameters of most radar systems can be divided into specific operating bands. Table 3-2 shows the primary bands in use today. As shown in the table, letter designations have been developed for most of the operating bands.

Radar frequencies have been selected to minimize atmospheric attenuation by rain and snow, clouds and fog, and (at some frequencies) by electrons in the air. The frequency bands must also support wide bandwidth radiation and high antenna gain.

3.3.2 Transmission Equipment

The operating parameters of a radar transmitter are entirely different from other transmitters discussed so far. Broadcast and satellite systems are characterized by medium power and continuous duty applications. Radar, on the other hand, is characterized by high power pulsed transmissions of relatively low duty cycle. The unique requirements of radar have led to the development of a technology that is foreign to most communications systems.

Improvements in semiconductor design and fabrication have made solid state radar sets possible. Systems producing several hundred Watts of output power at frequencies up to 2 GHz have been installed. Higher operating powers are achieved by using parallel amplification.

Despite inroads made by solid state devices, vacuum tubes continue to be the mainstay of radar technology. Tube-based systems consist of the following stages:

- Exciter. Generates the necessary RF and local-oscillator frequencies for the system.
- Power supply. Provides the needed operating voltages for the system.
- Modulator. Triggers the power output tube into operation. Pulse-shaping of the transmitted signal is performed in the modulator stage.
- RF amplifier. Converts the dc input from the power supply and the trigger signals from the modulator into a series of high energy, short duration pulse.

Exciter. The exciter stage of a radar set is comprised of an oscillator, frequency multiplier, and mixer. The signal produced depends on whether the transmitter output device operated is a power amplifier or an oscillator.

Transmitters using power oscillators such as *magnetrons* determine the RF frequency by the tuning of the device itself. In a conventional (*noncoherent*) radar system, the only frequency required is the *local oscillator* (LO). The LO differs from the magnetron frequency by an IF frequency, and this difference is usually maintained

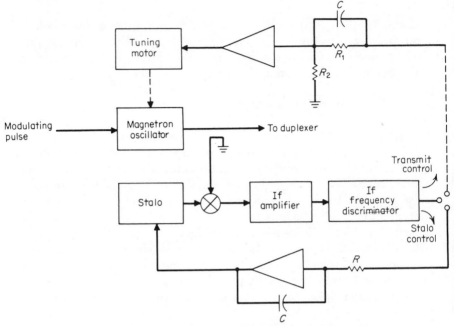

Figure 3-8 Two common approaches to automatic frequency control of a magnetron oscillator. (*Source:* Fink and Christiansen, *Electronics Engineers' Handbook, 3rd Ed.,* McGraw-Hill, New York, 1989.)

with an *automatic frequency control* (AFC) loop. Figure 3-8 shows a simple magnetron-based radar system with two methods of tuning: slaving the magnetron to follow the *stable local oscillator* (STALO), or slaving the STALO to follow the magnetron.

If the radar must use *coherent detection* (such as Doppler applications), a second oscillator, called a *coherent oscillator* (COHO), is required. The COHO operates at the IF frequency and provides a reference output for signal processing circuits.

The synchronizer circuit in the exciter supplies timing pulses to the various radar subsystems. In a simple marine radar this may consist of a single multivibrator that triggers the transmitter. In a larger system, 20 to 30 timing pulses may be required. These may turn on and off the beam current in various transmitter stages, start and stop RF pulse time attenuators, start display sweeps, and numerous other functions. Newer radar systems generate the required timing signals digitally. A digital synchronizer is illustrated in Figure 3-9.

Modulator. A radar system RF amplifier usually centers around one of two microwave devices: a *crossed-field tube* or *linear-beam*

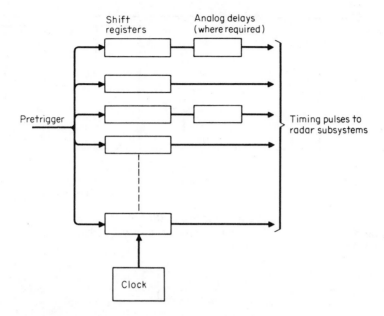

Shift registers

Analog delays (where required)

Pretrigger

Timing pulses to radar subsystems

Clock

Figure 3-9 Digital synchronizer system for radar applications. (Source: Fink and Christiansen, Electronics Engineers' Handbook, 3rd Ed., McGraw-Hill, New York, 1989.)

tube. Both are capable of high peak output power at microwave frequencies. To obtain high efficiency from a pulsed radar transmitter, it is necessary to cut off the current in the output tube between pulses. The modulator performs this function. Some RF tubes include control electrodes or grids to achieve the same result. Three common types of modulators are used today in radar equipment:

- *Line-type modulator* (Figure 3-10). This common radar modulator is used most often to pulse a magnetron. Between pulses, a charge is stored in a *pulse-forming network* (PFN). A trigger signal fires a thyratron tube, shorting the input to the PFN, which causes a voltage pulse to appear at the primary of transformer T1. The PFN components are chosen to produce a rectangular pulse at the magnetron cathode, with the proper voltage and current to excite the magnetron to oscillation. Advantages of this design include its simplicity. Drawbacks include the inability to electronically change the width of the transmitted pulse.
- *Active-switch modulator* (Figure 3-11). This system permits pulse width variation within the limitations of the energy stored in the

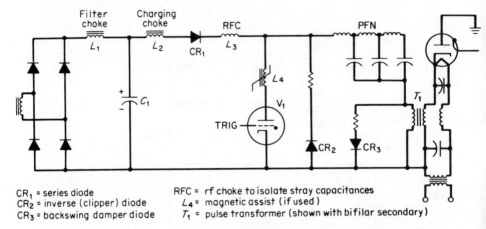

CR₁ = series diode

CR₂ = inverse (clipper) diode

CR₃ = backswing damper diode

RFC = rf choke to isolate stray capacitances

L_4 = magnetic assist (if used)

T_1 = pulse transformer (shown with bifilar secondary)

Figure 3-10 A line-type modulator for radar. (*Source:* Fink and Christiansen, *Electronics Engineers' Handbook, 3rd Ed.,* McGraw-Hill, New York, 1989.)

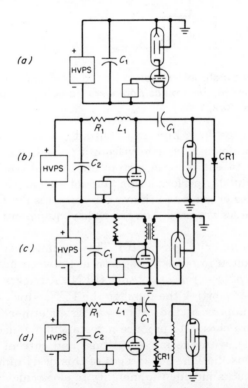

Figure 3-11 Active-switch modulator circuits: (a) direct-coupled system; (b) capacitor-coupled; (c) transformer-coupled; (d) capacitor- and transformer-coupled. (*Source:* Fink and Christiansen, *Electronics Engineers' Handbook, 3rd Ed.,* McGraw-Hill, New York, 1989.)

high voltage power supply. A switch tube controls the generation of RF by completing the circuit path from the output tube to the power supply, or by causing stored energy to be dumped to the output device. The figure shows the basic design of an active-switch modulator and three variations on the scheme. The circuits differ in their method of coupling power supply energy to the output tube (capacitor-coupled, transformer-coupled, or a combination of the two methods).

- *Magnetic modulator* (Figure 3-12). This design is the simplest of the thee modulators discussed. No thyratron or switching device is used. Operation of the modulator is based on the saturation characteristics of inductors L1, L2, and L3. A long-duration low-amplitude pulse is applied to L1, which charges C1. When C1 is charged, L2 saturates and the energy in C1 is transferred resonantly to C2. This process continues to the next stage (L3 and C3). The transfer time is set by selection of the components to be about one tenth that of the previous stage. At the end of the chain, a short-duration high-amplitude pulse is generated, exciting the RF output tube.

Antenna Systems. Because the applications for radar vary widely, so do antenna designs. Sizes range from less than 1 ft. to hundreds of feet in diameter. An antenna intended for radar applications must direct radiated power from the transmitter to the azimuth and elevation coordinates of the target. It must also serve as a receive antenna for the echo.

The coverage pattern of a typical antenna is shown in Figure 3-13. The diagram shows radiated field intensity as a function of azimuth and elevation. Because antenna patterns for transmit functions are

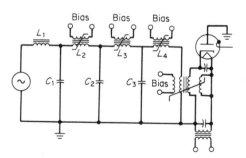

Figure 3-12 A magnetic modulator circuit. (*Source:* Fink and Christiansen, *Electronics Engineers' Handbook*, 3rd Ed., McGraw-Hill, New York, 1989.)

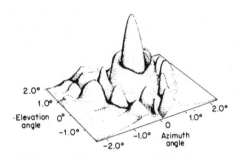

Figure 3-13 Typical coverage pattern of a radar antenna. The plot shows the 3-dimensional pencil-beam pattern of the AN/FPQ-6 radar antenna. (*Source:* D. D. Howard, Naval Research Laboratory in Fink and Christiansen, *Electronics Engineers' Handbook, 3rd Ed.,* McGraw-Hill, New York, 1989.)

identical to the receive pattern at the same frequency (the principle of *reciprocity*), the plot shown in the figure also describes the gain of the antenna as a receiving device.

There are three basic antenna designs for radar:

- *Search antenna.* Available in a wide variety of sizes, depending upon the application. Most conventional search antennas use mechanically scanned horn-feed reflectors. The horn radiates a spherical wavefront that illuminates the antenna reflector, the shape of which is designed to focus the radiated energy at infinity. The radiated beam is usually narrow in azimuth and wide in elevation (fan shaped).
- *Tracking antenna.* Intended primarily to make accurate range and angle measurements of the position of a particular target. Such antennas use circular apertures to form a pencil beam of about 1° in the X and Y coordinates. Operating frequencies in the S, C, and X bands are preferred because they allow a smaller aperture for the same transmitted beamwidth. The tracking antenna is physically smaller than most other types of comparable gain and directivity. This permits more accurate pointing at a given target. Figure 3-14 shows a typical tracking antenna for ground-based applications.
- *Multifunction array.* An electrically-steered antenna used for both airborne and ground-based applications. An array antenna consists of individual radiating elements that are driven together to produce a plane wavefront in front of the antenna aperture. Most ar-

Figure 3-14 A conventional tracking radar antenna, the AN/TPN-19, used for airport surveillance. (*Source:* Raytheon Co. in Fink and Christiansen, *Electronics Engineers' Handbook, 2rd Ed.,* McGraw-Hill, New York, 1982

rays are flat with the radiating elements spaced about 0.6 wavelength apart. Steering is accomplished by changing the phase relationships of groups of radiating elements with respect to the array.

3.3.3 Phased Array Antennas

Phased array antennas are steered by tilting the phase front independently in two orthogonal directions called the *array coordinates.*

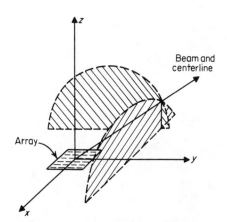

Figure 3-15 Beam-steering contours for a planar array antenna. (*Source:* Fink and Christiansen, *Electronics Engineers' Handbook, 3rd Ed.,* McGraw-Hill, New York, 1989.)

Scanning in either array coordinate causes the beam to move along a cone whose center is at the center of the array. The paths the beam follows when steered in the array coordinates are illustrated in Figure 3-15 (the Z axis is *normal* to the array). As the beam is steered away from the array normal, the projected aperture in the beam's direction varies, causing the beamwidth to vary proportionately.

Arrays can be classified as either active or passive. Active arrays contain duplexers and amplifiers behind every element or group of elements of the array. Passive arrays are driven from a single feed point. Active arrays are capable of high power operation, higher in fact than a conventional radar antenna.

Both passive and active arrays must divide the signal from a single transmission line among all the elements of the system. This can be accomplished through one of the following methods:

• *Optical feed.* A single feed, usually a monopulse horn, is used to illuminate the array with a spherical phase front (see Figure 3-16). Power collected by the rear elements of the array is transmitted through the phase shifters to produce a planar front and steer the array. The energy may then be radiated from the other side of the array or reflected and reradiated through the collecting elements. In the later case, the array acts as a steerable reflector.

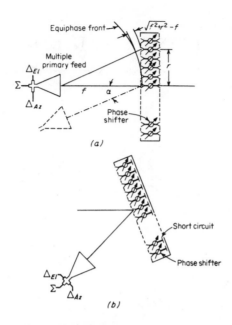

Figure 3-16 Optical antenna feed systems: (a) lens; (b) reflector. (*Source: Fink and Christiansen, Electronics Engineers' Handbook, 3rd Ed., McGraw-Hill, New York, 1989.*)

- *Corporate feed.* May utilize a series-feed network (Figure 3-17) or parallel-feed network (Figure 3-18). Both designs use transmission-line components to divide the signal among the elements. Phase shifters can be located at the elements or within the dividing network. Both the series- and parallel-feed systems have several variations, as shown in the figures.
- *Multiple-beam network.* Capable of forming simultaneous beams with a given array. The *Butler matrix,* shown in Figure 3-19, is one such technique. It connects the N elements of a linear array to N feed points corresponding to N beam outputs. The phase shifter is one of the most critical components of the system. It produces controllable phase shift over the operating band of the array. Digital and analog phase shifters have been developed using both ferrites and *pin diodes.*

Frequency scan is another type of multiple-beam network, but one that does not require phase shifters, dividers or beam-steering computers. Element signals are coupled from points along a transmission line. The electrical path length between elements is longer than the

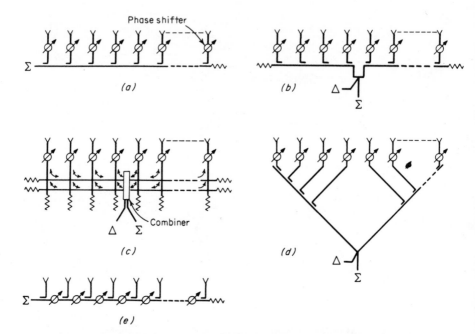

Figure 3-17 Series-feed networks: (a) end feed; (b) center feed; (c) separate optimization; (d) equal path length; (e) series phase shifters. (*Source:* Fink and Christiansen, *Electronics Engineers' Handbook, 3rd Ed.,* McGraw-Hill, New York, 1989.)

physical separation, and a small frequency change will cause a phase change between elements large enough to steer the beam. This technique can be applied only to one array coordinate. If a two dimensional array is required, phase shifters are normally used to scan the other coordinate.

Phase Shift Devices. The design of a phase-shifter must meet two primary criteria: low transmission loss and high power handling capability. The *Reggia-Spencer* phase shifter meets both requirements. The device consists of a ferrite rod mounted inside a waveguide (Figure 3-20) that delays the RF signal passing through the waveguide, permitting the array to be steered. The amount of phase shift can be controlled by the current in the solenoid, because of the effect a magnetic field has on the permeability of the ferrite. This design is a *reciprocal phase shifter,* meaning that the device exhibits

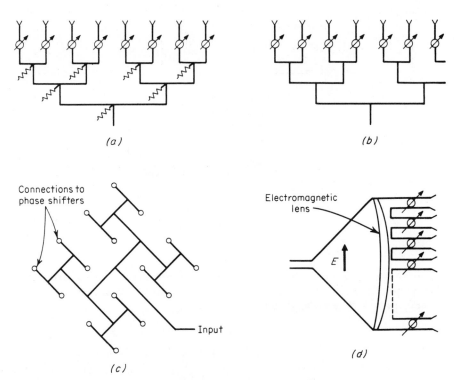

Figure 3-18 Parallel-feed networks: (a) matched corporate feed; (b) reactive corporate feed; (c) reactive stripline; (d) multiple reactive divider. (*Source:* Fink and Christiansen, *Electronics Engineers' Handbook, 3rd Ed.,* McGraw-Hill, New York, 1989.)

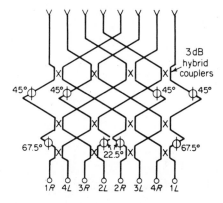

Figure 3-19 The Butler matrix beam-forming network. (*Source:* Fink and Christiansen, *Electronics Engineers' Handbook, 3rd Ed.,* McGraw-Hill, New York, 1989.)

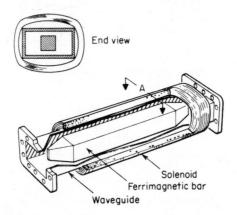

Figure 3-20 Basic concept of a Reggia-Spencer phase shifter. (*Source:* Fink and Christiansen, *Electronics Engineers' Handbook, 3rd Ed.*, McGraw-Hill, New York, 1989.)

the same phase shift for signals passing in either direction. *Nonreciprocal phase shifters,* where phase-shift polarity reverses with the direction of propagation, are also available.

Phase shifters have also been developed using pin diodes in transmission line networks. One configuration, shown in Figure 3-21, uses diodes as switches to change the signal path length of the network. A second type uses pin diodes as switches to connect reactive loads

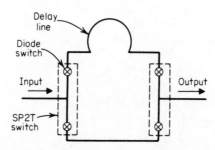

Figure 3-21 Switched-line phase shifter using pin diodes. (*Source:* Fink and Christiansen, *Electronics Engineers' Handbook, 3rd Ed.*, McGraw-Hill, New York, 1989.)

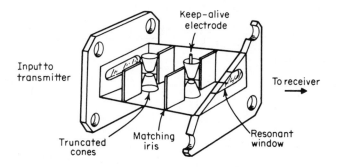

Figure 3-22 Typical construction of a TR tube. (*Source:* Fink and Christiansen, *Electronics Engineers' Handbook, 3rd Ed.*, McGraw-Hill, New York, 1989.)

across a transmission line. When equal loads are connected with a quarter-wave separation, a pure phase shift results.

3.3.4   Microwave Components

The radar transmitter, antenna, and receiver are all connected through RF transmission lines to a *duplexer*. The duplexer acts as a switch connecting the transmitter to the antenna while radiating, and the receiver to the antenna while listening for echoes. Filters, receiver protection components, and rotary RF joints may also be part of this hardware.

Duplexer. The duplexer is an essential component of any radar system. The switching elements used in a duplexer include gas tubes, ferrite circulators, and pin diodes. Gas tubes are the most common. A typical gas-filled *TR tube* is shown in Figure 3-22. Low power RF signals pass through the tube with very little attenuation. Higher power signals, however, cause the gas to ionize and present a short circuit to the RF energy.

Figure 3-23 illustrates a *balanced duplexer* using hybrid junctions and TR tubes. When the transmitter is on, the TR tubes fire and reflect the RF power to the antenna port of the input hybrid. During the receive portion of the radar function, signals picked up by the antenna are passed through the TR tubes and on to the receiver port of the output hybrid.

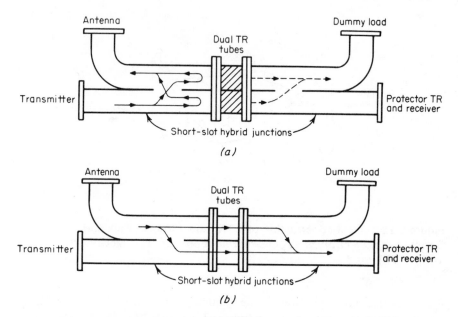

Figure 3-23 Balanced duplexer circuit using dual TR tubes and two short-slot hybrid junctions: (a) transmit mode; (b) receive mode. (*Source:* Fink and Christiansen, *Electronics Engineers' Handbook, 3rd Ed.*, McGraw-Hill, New York, 1989.)

Circulators and Diode Duplexers. Newer radar systems often use a ferrite circulator as the duplexer. A TR tube is required in the receiver line to protect input circuits from transmitter power reflected by the antenna because of an imperfect match. A 4-port circulator is generally used with a load between the transmitter and receiver ports so that power reflected by the TR tube is properly terminated.

Pin diode switches have also been used in duplexers to perform the protective switching function of TR tubes. Pin diodes are more easily applied in coaxial circuitry, and at lower microwave frequencies. Multiple diodes are used when a single diode cannot withstand the expected power.

Filters. Microwave filters are sometimes used in the transmit path of a radar system to suppress spurious radiation, or in the receive signal path to suppress spurious interference.

Harmonic filters are commonly used in the transmission chain to absorb harmonic energy generated by the system, preventing it from being radiated or reflected back from the antenna. Figure 3-24 shows

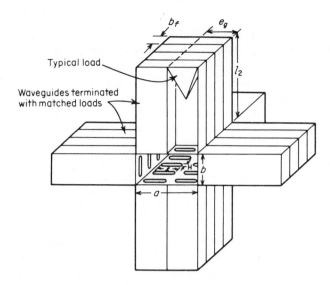

Figure 3-24 Construction of a dissipative waveguide filter. (*Source:* Fink and Christiansen, *Electronics Engineers' Handbook, 3rd Ed.*, McGraw-Hill, New York, 1989.)

a filter in which harmonic energy is coupled out through holes in the walls of the waveguide to matched loads.

Narrow-band filters in the receive path, often called *preselectors,* are built using mechanically tuned cavity resonators or electrically-tuned *TIG resonators.* Preselectors can provide up to 80 dB suppression of signals from other radar transmitters in the same RF band, but at a different operation frequency.

3.4 Electronic Navigation Systems

Navigation systems based on radio transmissions are used every day by commercial airlines, general-aviation aircraft, ships, and the military. Electronic position-fixing systems are also used in surveying work. While the known propagation speed of radio waves allows high accuracies to be obtained in free space, multipath effects along the surface of the Earth are the primary enemies of practical airborne and shipborne systems. A number of different navigation tools, therefore, have evolved to provide the needed accuracy and coverage area.

Electronic navigation systems can be divided into three primary categories:

- Long-range. Useful for distances of greater than 200 mi., long-range systems are primarily used for transoceanic navigation.
- Medium-range. Useful for distances of 20 to 200 mi., medium-range systems are mainly used in coastal areas and above populated land masses.
- Short-range. Useful for distances of less than 20 mi., short-range systems are used for approach, docking, or landing applications.

Electronic navigation systems can be further divided into *cooperative* or *self-contained.* Cooperative systems depend on transmission, one- or two-way, between one or more ground stations and the vehicle. Such systems are capable of providing the vehicle with a location fix, independent of its previous position. Self-contained systems are contained entirely in the vehicle and may be radiating or nonradiating. In general, they measure the distance traveled, and have errors that increase with time or distance. The type of system chosen for a particular application depends upon a number of considerations, including how often the location of the vehicle must be determined and the accuracy required.

Because aircrafts and ships may travel to any part of the world, many electronic navigation systems have received standardization on an international scale.

Virtually all radio frequencies have been used in navigation at one point or another. Systems operating at low frequencies typically use high power transmitters with massive antenna systems. With a few exceptions, frequencies and technologies have been chosen to avoid dependence on ionospheric reflection. Such reflections can be valuable in communications systems, but are usually unpredictable. Table 3-3 lists the principal frequency bands used for radionavigation.

3.4.1   Direction Finding

Direction finding (DF) is the oldest and most widely used navigation aid. The position of a transmitter may be determined by comparing the arrival coordinates of the radiated energy at two or more known points. Conversely, the position of a receiving point may be determined by comparing the direction coordinates from two or more known transmitters.

The weakness of this system is its susceptibility to site errors. The chief weapon against error is the use of a large DF antenna aperture. In many cases, a multiplicity of antennas, suitably combined,

Table 3-3 Radio frequencies used for electronic navigation. (*Source:* Fink and Christiansen, *Electronics Engineers' Handbook, 3rd Ed.,* McGraw-Hill, New York, 1989.)

System	Frequency band	No. of stations	No. of vehicles
Omega	10–13 kHz	8	10,000
VLF Comm.	16–24 kHz	10	5,000
Decca	70–130 kHz	150	30,000
Loran-C/D	100 kHz	50	10,000
Lf range	200–400 kHz	*	*
ADF/NDB†	200–1,600 kHz	4,000	106,000
Coastal DF†	285–325 kHz	1,000	100,000
Consol	250–350 kHz	15	5,000
Loran A	2 MHz	*	*
Marker beacon†	75 MHz	2,500	150,000
ILS localizer†	108–112 MHz	1,200	150,000
VOR†	108–118 MHz	2,000	250,000
ILS glide slope†	329–335 MHz	1,200	150,000
Transit	150,400 MHz	6	5,000
DME†, Tacan	960–1,21 MHz	2,000	70,000
ATCRBS†	1,030, 1,090 MHz	800	250,000
GPS	1,227, 1,575 MHz	‡	‡
Altimeter	4,200 MHz	. . .	5,000
Talking beacons	9 GHz	3	1,000
MLS	5 GHz	‡	‡
Weather radar	5, 9 GHz	. . .	10,000
Doppler radar	10–20 GHz	. . .	5,000

* Obsolescent.
† Internationally standardized systems.
‡ In development.

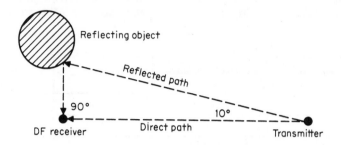

Figure 3-25 Direction finding error resulting from beacon reflections.
(*Source:* Fink and Christiansen, *Electronics Engineers' Handbook, 3rd Ed.,*
McGraw-Hill, New York, 1989.)

can be made to favor the direct path and discriminate against indirect paths (see Figure 3-25).

Ship navigation is a common application of DF. Coastal beacons operate in the 285–325 kHz band specifically for ship navigation. This low frequency provides ground-wave coverage over seawater to about 1,000 mi. Operating powers vary from 100 W to 10 kW. A well-designed shipboard DF system can provide accuracies of about ±2° under typical conditions.

3.4.2   Two-Way Distance Ranging

By placing a transponder on a given target, automatic distance measuring can be accomplished, as illustrated in Figure 3-26. The system receives an interrogator pulse and replies to it with another pulse, usually on a different frequency. Various codes can be employed to limit responses to a single target or class of target.

Distance-measuring equipment (DME) systems is one application of two-way distance ranging. An airborne interrogator transmits 1 kW pulses at a 30 Hz rate on one of 126 channels spaced 1 MHz apart. (The operating band is 1.025 to 1.150 GHz). A ground transponder responds with similar pulses on another channel at 63 MHz above or below the interrogating channel.

In the airborne set, the received signal is compared with the transmitted signal, their time difference derived, and a direct digital reading of miles is displayed with a typical accuracy of ±0.2 mi.

Ground transponders are arranged to handle interrogation for up to 100 aircraft simultaneously.

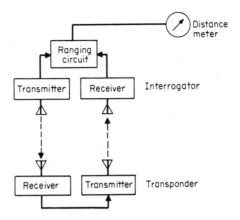

Figure 3-26 The concept of two-way distance ranging. (*Source:* Fink and Christiansen, *Electronics Engineers' Handbook, 3rd Ed.,* McGraw-Hill, New York, 1989.)

3.4.3   Differential Distance Ranging

Two-way ranging requires a transmitter at both ends of the link. The differential distance ranging system avoids carrying a transmitter on the vehicle by placing two on the ground. One is a master, and the other a slave repeating the master (see Figure 3-27). The receiver measures the difference in the arrival of the two signals. For each time difference, there is a *hyperbolic line of position* that defines the target location. (Such systems are known as *hyperbolic* systems.) The transmissions may be either pulsed or continuous-wave using different carrier frequencies. At least two pairs of stations are needed to produce a fix.

If both stations in a differential distance ranging system are provided with stable, synchronized clocks, distance measurements can be accomplished through one-way transmissions whose elapsed time is measured with reference to the clocks. This mode of operation is referred to as *one-way distance ranging*. The concept is illustrated in Figure 3-28.

Loran C. Hyperbolic positioning is used in the *Loran C* navigation system. Chains of transmitters, located along coastal waters, radiate pulses at a carrier frequency of 100 kHz. Because all stations operate on the same frequency, discrimination between chains is accomplished by different pulse-repetition frequencies. A typical chain con-

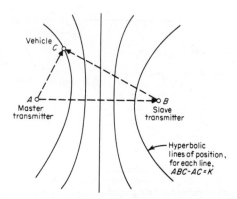

Figure 3-27 The concept of differential distance ranging (hyperbolic).
(*Source:* Fink and Christiansen, *Electronics Engineers' Handbook, 3rd Ed.,*
McGraw-Hill, New York, 1989.)

sists of a master station and two slaves, about 600 miles from the
master. Each antenna is 1,300 ft. high and is fed 5 MW pulses,
which build up to peak amplitude in about 50 μs and then decay to
zero in approximately 100 μs. The slow rise and decay times are
necessary to keep the radiated spectrum within the assigned band
limits of 90 to 100 kHz.

To obtain greater average power at the receiver without resorting
to higher peak power, the master station transmits groups of 9
pulses, 1 ms apart. These groups are repeated at rates ranging from
10 to 25 per second. Within each pulse, the RF phase can be varied
for communications purposes.

Coverage of Loran C extends to all U.S. coastal areas, plus certain
areas of the North Pacific, North Atlantic, and Mediterranean. There
are currently 17 chains employing about 50 transmitters.

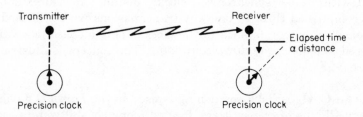

**Figure 3-28 The concept of one-way distance ranging. (*Source:* Fink and
Christiansen, *Electronics Engineers' Handbook, 3rd Ed.,* McGraw-Hill, New
York, 1989.)**

Omega. Omega is another navigation system based on the hyperbolic concept. The system is designed to provide worldwide coverage from just eight stations. Omega operates on the VLF band, from 10–13 kHz. At this low frequency, skywave propagation is relatively stable. Overall accuracy is on the order of 1 mi., even at ranges of 5,000 mi.

There are no masters or slaves; each station transmits according to its own standard. Each station has its own operating code and transmits on one frequency at a time for a minimum of about 1 s. The cycle is repeated every 10 s. These slow rates are necessary because of the high Qs of the transmitting antennas.

A simple Omega receiver monitors for signals at 10.2 kHz and compares emissions from one station against those of another by use of an internal oscillator. The phase difference data are transferred to a map with hyperbolic coordinates.

Most Omega receivers are also able to use VLF communications stations for navigation. There are about ten such facilities operating between 16 and 24 kHz. Output powers range from 50 kW to 1 MW. Frequency stability is maintained to 1 part in 10^{12}. This allows one-way DME to be accomplished with a high degree of accuracy.

3.5 Microwave Radio

Analog microwave radio relay systems carry the bulk of long-haul telecommunications in the U.S. and most other countries. The major common-carrier bands and their applications are shown in Table 3-4. The primary goals of microwave relay technology have been to increase channel capacity and lower costs. Solid state devices have provided the means to accomplish these goals. Current efforts focus on the use of fiber optic land-lines for terrestrial long-haul communications systems. Satellite circuits have also been used extensively for long-haul common-carrier applications.

Single-sideband amplitude modulation is used for microwave systems because of its spectrum efficiency. Single-sideband systems, however, require a high degree of linearity in amplifying circuits. Several techniques have been used to provide the needed channel linearity. The most popular is *amplitude predistortion* to cancel the inherent nonlinearity of the power amplifier.

The final stage of frequency division multiplexing (FDM), which might, for example, combine three 600 channel master groups, is often located some distance (a mile or more) from the transmitter site. A coaxial wire-line is used to bring the baseband signal to the microwave equipment. Depending on the distance, intermediate re-

Table 3-4 Common-carrier microwave frequencies used in the U.S.
(*Source:* Fink and Christiansen, *Electronics Engineers' Handbook, 3rd Ed.,*
McGraw-Hill, New York, 1989.)

Band, GHz	Allotted frequencies, MHz	Bandwidth, MHz	Application
2	2,110–2,130 2,160–2,180	20	Limited
4	3,700–4,200	20	Major long-haul micro-wave relay band
6	5,925–6,425	500	Long and short haul
11	10,700–11,700	500	Short haul
18	17,700–19,700	1,000	Short haul, limited use
30	27,500–29,500	2,000	Short haul, experimental

peaters may be used. The baseband signal is applied to an FM terminal (FMT) which frequency modulates a carrier of typically 70 MHz. This IF signal then modulates a 20 MHz-wide channel in the 4 GHz band.

3.6 Induction Heating

Induction heating is achieved by placing a coil-carrying alternating current, adjacent to a metal workpiece so that the magnetic flux produced induces a voltage in the workpiece. This causes current flow and heats the workpiece. Power sources for induction heating include:

- Motor-generator sets, which operate at low frequencies and provide outputs from 1 kW to more than 1 MW.
- Vacuum-tube oscillators, which operate at 3 kHz to several hundred MHz at power levels of 1 kW to several hundred kilowatts. (Figure 3-29 shows a 20 kW induction heater using a vacuum tube as the power generating device.)
- Inverters, which operate at 10 kHz or more at power levels of as much as several megawatts. Inverters utilizing thyristors (silicon controlled rectifiers) are replacing motor-generator sets in high power applications.

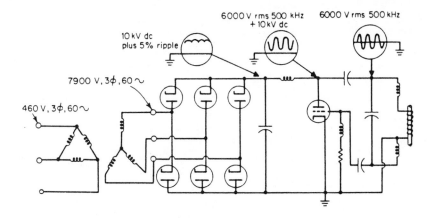

Figure 3-29 20 kW induction heater circuit. (*Source:* Fink and Christiansen, *Electronics Engineers' Handbook, 3rd Ed.,* McGraw-Hill, New York, 1989.)

3.6.1 Dielectric Heating

Dielectric heating is a related application for RF technology. Instead of heating a conductor, as in induction heating, dielectric heating relies on the capacitor principle to heat an insulating material. The material to be heated forms the dielectric of a capacitor, to which power is applied. The heat generated is proportional to the *loss factor* (the product of the dielectric constant and the power factor) of the material. Because the power factor of most dielectrics is low at low frequencies, the range of frequencies employed for dielectric heating is higher than for induction heating. Frequencies of a few megahertz to several gigahertz are common.

Bibliography

1. Benson, K., and J. Whitaker, *Television and Audio Handbook for Technicians and Engineers,* McGraw-Hill, New York, 1989.
2. Fink, D., and D. Christiansen, *Electronics Engineer's Handbook,* Third Edition, McGraw-Hill, New York, 1989.
3. Fink, D., and D. Christiansen, *Electronics Engineer's Handbook,* Second Edition, McGraw-Hill, New York, 1982.
4. Jordan, Edward C., *Reference Data for Engineers: Radio, Electronics, Computer and Communications,* Seventh Edition, Howard W. Sams Company, Indianapolis, IN, 1985.

Solid State RF Devices

4.1 Introduction[1]

Solid state devices play an increasingly important role in the generation of RF energy. Designs based on semiconductors offer a number
of advantages over vacuum tubes, including:

* Reduced size and weight
* Lower operating voltages
* No warm-up period required
* Practical fault-tolerant designs
* Simplified cooling
* Improved operating efficiency (depending on the frequency, power
  level, and type of modulation)
* Reduced susceptibility to mechanical shock and vibration

As of this writing, solid state devices are routinely used at power
levels of 250 W (CW). As designers find new ways to improve operating efficiency and remove heat generated during use, the maximum
operating power will rise. It is unlikely, however, that semiconductors will be able to operate in a CW mode in excess of 500 W cost-ef-

---

1 Portions of this chapter were adapted from *Television and Audio Handbook for Technicians and Engineers,* K. Blair Benson and Jerry Whitaker
co-authors, McGraw-Hill, 1990.

fectively anytime soon. In order to achieve greater output levels, devices — indeed entire amplifiers — must be operated in parallel. Figure 4-1 illustrates an RF amplifier using both parallel devices and parallel amplifiers. The input signal is split to drive each amplifier, and the outputs are combined to feed the load.

Parallel amplification is attractive from several standpoints. First, redundancy is a part of the basic design. If one device or one amplifier fails, the remainder of the system will continue to operate. Second, lower-cost devices can be used. It is often less expensive to put two 100 W transistors into a circuit than it is to put one 200 W device. Third, troubleshooting the system is simplified because an entire RF module can be substituted to return the system to operation. The defective part can be replaced at a later time by the technician, or returned to the factory for repairs.

Solid state systems are not, however, without their drawbacks. A high power transmitter using vacuum tubes is much simpler in design than a comparable solid state system. The greater the number of parts, the higher the potential for system failure. Higher parts counts usually translate to higher overall failure rates. It is only fair to point out, however, that failures in a parallel, fault-tolerant design will usually not cause the entire system to fail. Instead, some parameter, typically peak output power, will drop when one or more amplifier modules is out of commission.

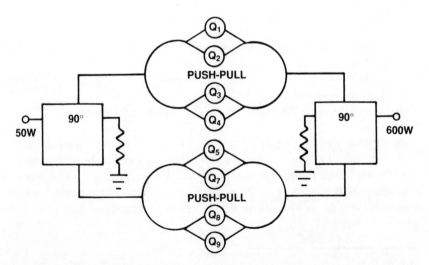

Figure 4-1 Schematic diagram of a 600 W VHF amplifier using eight FETs in a parallel device/parallel module configuration. *(Source: Broadcast Engineering Magazine.)*

This discussion assumes that the design of a solid state system is truly fault-tolerant. For a system to provide the benefits of parallel design, power supplies, RF divider and combiner networks, and supervisory/control systems, must also be capable of independent operation.

4.1.1   Operating Efficiency

The efficiency of a solid state transmitter may or may not be better than a tube transmitter of the same operating power and frequency. Much depends on the type of modulation used and the frequency of operation. For example, new solid state designs have led to significant improvements in the operating efficiency of high power (50 kW) AM broadcast transmitters. This improvement has come from both improved devices and new modulation schemes. High power (30 to 60 kW) solid state television transmitters, on the other hand, operate with about the same overall efficiency as their vacuum tube counterparts.

Choosing between a solid state or vacuum tube design is not as simple as it might appear on the surface. Many items must be considered, and tradeoffs accepted.

4.2   Bipolar Transistors

A bipolar transistor has two PN junctions that act exactly like a simple diode PN junction: the *base-emitter junction* and the *base-collector junction* (Figure 4-2 shows an NPN junction transistor). In typical use, the first junction is *forward biased,* causing normal conduction, and the second junction is *reversed bias.* If the material of the base was thick, the flow of electrons into the p-material base junction of an NPN transistor would go entirely into the base-emitter junction and no current would flow in the reverse-biased collector-base junction.

If, however, the base junction was thin, electrons would diffuse in the semiconductor crystal lattice and move into the base-collector junction, having been injected into the base material of the base-emitter junction. This diffusion occurs because excess electrons moving into one location bump electrons in adjacent semiconductor molecules, which bump other molecules, and so on. Thus, a collector current would flow nearly as large as the injected emitter current.

The ratio of collector to emitter current is referred to as *alpha* (α), or the common-base current gain of the transistor. Alpha for most transistors is slightly less than 1.0. The portion of emitter current

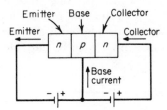

Figure 4-2 An NPN junction transistor. (*Source:* Fink and Christiansen, *Electronics Engineers' Handbook, 3rd Ed.,* McGraw-Hill, New York, 1989.)

not flowing into the collector will flow as a base current in the same direction as the collector current. The ratio of collector current to base current is referred to as *beta* (β). Beta is the conventional (common-emitter) current gain of the transistor. It may be as low as 5 in power transistors operating at maximum current levels, or as high as 5,000 in super-beta transistors operating in the region of maximum current gain.

4.2.1   NPN and PNP Transistors

Bipolar transistors are identified by the sequence of semiconductor material in an emitter-to-collector sequence. NPN transistors operate normally with a positive voltage on the collector with respect to the emitter. PNP transistors operate normally with a negative voltage at the collector with respect to the emitter. The flow of current in a PNP device is primarily the result of *holes* (absent excess electrons) in the crystal lattice at locations of current flow.

Because the diffusion velocity of holes is slower than electrons, PNP transistors have more junction capacitance and slower speed than NPN transistors of the same size. Holes in PN junctions are *minority carriers* of electric current as opposed to electrons, which are *majority carriers*. Holes and electrons can move freely in resistors or in the conductive channel of field-effect transistors.

Figure 4-3 shows the basic symbols used to describe transistor parameters.

Transistor Materials. Silicon is the most common transistor material used today. Silicon permits transistor junction temperatures as high as 200°C. The nominal base-emitter voltage is about 0.7 V, and collector-emitter voltage ratings of up to hundreds of volts are available. At room temperature silicon-based transistors can dissipate hundreds of watts with proper heat removal.

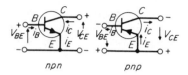

Figure 4-3 Transistor symbols for NPN and PNP devices. (*Source:* Fink and Christiansen, *Electronics Engineers' Handbook, 3rd Ed.,* McGraw-Hill, New York, 1989.)

In the early 1980s, transistors made of gallium arsenide (GaS) and similar materials became available for use in microwave and high speed circuits. Such devices take advantage of the high diffusion speed and low capacitance of GaS. However, difficulty in fabricating such devices has limited their use to specialized applications.

Basic Configurations. Transistors are used in a wide variety of circuit designs. Although operation of each system is unique, all are based on one of three fundamental configurations:

- *Common emitter,* in which the emitter is grounded (the most common design). The voltage source supplies current only to the base. Base current is the difference between the emitter and collector currents; it is much smaller than either. Hence, current gain (Ic/Ib) and input impedance is high.
- *Common base,* in which the base is grounded. Current gain is slightly less than one. Voltage and power amplification can be achieved because the output impedance is much higher than the input impedance.
- *Common collector,* in which the collector is grounded. The source voltage and the output voltage are in series, and have opposing polarities. This negative-feedback arrangement yields a high input impedance and approximately unity voltage gain. Current gain is about the same as the common-emitter connection.

Figure 4-4 illustrates the three configurations. Table 4-1 lists the primary operating characteristics of the three circuits.

The common-emitter, common-base, and common-collector are roughly analogous to the grounded-cathode, grounded-grid, and grounded-plate (cathode-follower) connections, respectively, of the vacuum tube.

Table 4-1 Basic bipolar device amplifier configurations and operating characteristics. (*Source:* Rohde and Bucher, *Communications Receivers: Principles and Design,* McGraw-Hill, New York, 1988.)

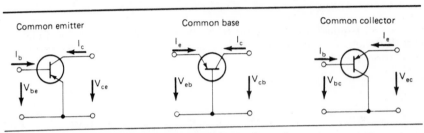

| | Characteristics of basic configurations | | |
	Common emitter	Common base	Common collector
Input impedance Z_1	Medium Z_{1e}	Low $Z_{1b} \approx \dfrac{Z_{1e}}{h_{fe}}$	High $Z_{1c} \approx h_{fe} R_L$
Output impedance Z_2	High Z_{2e}	Very high $Z_{2b} \approx Z_{2e} h_{fe}$	Low $Z_{2c} \approx \dfrac{Z_{1e} + R_g}{h_{fe}}$
Small-signal current gain	High h_{fe}	< 1 $h_{fb} \approx \dfrac{h_{fe}}{h_{fe} + 1}$	High $\gamma \approx h_{fe} + 1$
Voltage gain	High	High	< 1
Power gain	Very high	High	Medium
Cutoff frequency	Low f_{hfe}	High $f_{hfb} \approx h_{fe} f_{hfc}$	Low $f_{hfc} \approx f_{hfe}$

4.2.2   Volt-Ampere Characteristics

The performance of a transistor over wide ranges of current and voltage is determined from static characteristic curves. Figure 4-5 shows an example curve for a device in a common-emitter configuration.

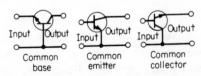

Figure 4-4 Basic circuit connections for an NPN transistor. (*Source:* Fink and Christiansen, *Electronics Engineers' Handbook, 3rd Ed.,* McGraw-Hill, New York, 1989.)

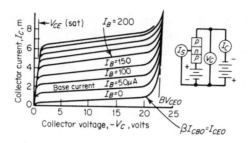

Figure 4-5 Collector voltage/current characteristic curves for a common-emitter bipolar transistor. (*Source:* Fink and Christiansen, *Electronics Engineers' Handbook, 3rd Ed.,* McGraw-Hill, New York, 1989.)

Collector current (I_c) is plotted as a function of collector-to-emitter voltage (V_c) for constant values of base current (I_b). Maximum collector voltage for the common-emitter configuration is limited by either *punch-through* or *avalanche breakdown,* whichever is lower. When a critical electric field is reached, avalanche occurs as a result of intensive current multiplication; current increases rapidly with little increase in voltage. (See Section 4.4.3.)

The grounded-emitter saturation voltage [$V_{CE(sat)}$] is also shown in the transistor's characteristic curves. This parameter is especially important in switching applications.

Dynamic variations of voltage and current are analyzed by a *load line* on the characteristic curves. For a linear transistor with load resistance R_l, the output varies along a load line of slope-$1/R_l$ about the dc operating point (see Figure 4-6). The load line for a common-emitter device in a switching application is shown in Figure 4-7. When the base current is zero, the collector is effectively an open

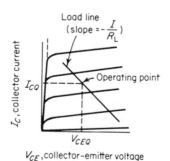

V_{CE}, collector-emitter voltage

Figure 4-6 Load line curves for a linear transistor amplifier. (*Source:* Fink and Christiansen, *Electronics Engineers' Handbook, 3rd Ed.,* McGraw-Hill, New York, 1989.)

circuit and only leakage current flows in the collector. The device is turned on by applying base current, which decreases the collector voltage to the *saturation value*.

4.2.3   Transistor Impedance and Gain

When analyzing the operating limits of a transistor, it is useful to first assume that the device will behave as an ideal transistor, and then examine degradations due to nonideal behavior. The operating impedance and gain are normally referenced to the common-emitter connection, which also results in the highest gain. If it is assumed that the transistor has a fixed current gain, then collector current will be equal to base current multiplied by the current gain of the transistor. Emitter current will be the sum of both the base and collector currents. Because the collector-base junction is reverse-biased, the output impedance of the ideal transistor is very high.

Practical bipolar transistors depart from the ideal model. Each terminal of the device exhibits a combination of series resistance and inductance, and shunt capacitance. These factors cause the transistor to have lower gain than predicted. In addition, practical transistors exhibit some amount of inductance and capacitance *between* terminals that cause a further reduction in available gain, particularly at low currents and with high load resistances. Transistor capacitances are caused in part by the finite diffusion velocities in silicon, and in part by the physical structure of the device. These stray elements cause transistor current gain to decrease with increasing frequency. At some point the transistor reaches unity current gain, identified as the *transition frequency* (see Figure 4-8). Stray elements

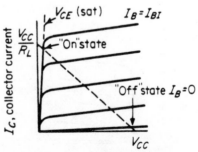

V_{CE}, collector-emitter voltage

Figure 4-7 Load line curves for a common-emitter switching circuit. (*Source:* **Fink and Christiansen,** *Electronics Engineers' Handbook, 3rd Ed.,* **McGraw-Hill, New York, 1989.)**

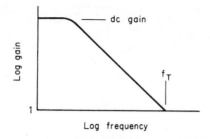

Figure 4-8 Amplitude-frequency response of a common-emitter amplifier. (*Source:* K. Blair Benson, *Audio Engineering Handbook,* McGraw-Hill, New York, 1988.)

also result in a feedback current from collector to base through the base-collector capacitance. An energy storage phenomenon may also be observed, similar to the energy storage effect in a rectifier diode. Stray capacitances at the collector will require a finite time to charge and discharge, limiting the speed at which the device may be switched on and off. This is an important factor in the design of an RF amplifier. Figure 4-9 charts transistor power output as a function of operating frequency for both pulsed and CW operation.

4.2.4   Noise

Every resistor creates noise with equal and constant energy for each hertz of bandwidth, regardless of frequency. It is useful to note that a 1 k Ω resistor at room temperature has an open-circuit output noise voltage of 4 nanovolts per *root-hertz.* This converts to 40 nV for a 100 Hz bandwidth, or 400 µV for a 10 MHz bandwidth.

Bipolar transistors also create noise in their input and output circuits. Transistor noise is effectively generated at the input junction; all noise ratings are referenced to the input junction.

In an ideal bipolar transistor, the voltage noise at the base is equivalent to the voltage noise of a resistor whose value is twice the transistor input conductance at the emitter terminal. The current noise is equivalent to the noise generated by a resistor whose value is twice the transistor input conductance at the input terminal.

Practical transistors do not behave in an ideal fashion. All transistors exhibit voltage and current noise energy that decreases with frequency. At some *corner frequency* noise becomes independent of frequency. Very low noise transistors may have a corner frequency as low as a few hertz; ordinary high-frequency devices may have a corner frequency well above the audio frequency range.

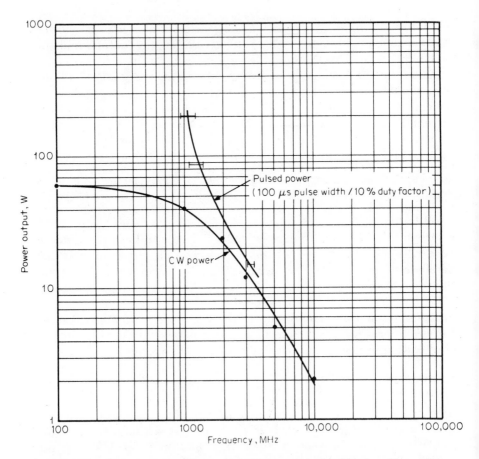

Figure 4-9 Power output as a function of frequency for bipolar silicon RF transistors. (*Courtesy of R. A. Gilson in Fink and Christiansen, Electronics Engineers' Handbook, 3rd Ed.,* McGraw-Hill, New York, 1989.)

Transistor noise may also be increased by operating a device too close to its maximum current rating. Poor transistor design or manufacturing errors may result in transistors that exhibit *popcorn* noise, so named after the audible characteristics of a random low-level switching effect.

4.2.5   Large Signal Characteristics

Transistor analysis thus far has dealt only with small signal behavior. When a transistor has to handle large currents, other limitations must be observed.

When handling low-frequency signals, a transistor may be viewed as a variable-controlled resistor between the supply voltage and the load impedance. The *quiescent operating point* (the absence of an input ac signal) is chosen for maximum symmetrical signal excursions (both positive and negative directions). This operating point is critical in class B push-pull amplifiers where one transistor conducts current to the load during the "positive" part of the cycle, and the other conducts during the "negative" part.

Other large signal limitations include the maximum power dissipation capability of the device under worst-case conditions of supply voltage, load impedance, drive signal, and ambient temperature.

Safe Operating Area. The safe operating area (SOA) of a power transistor is the single most important parameter in the design of a solid-state amplifier. Fortunately, advances in diffusion technology, masking, and device geometry have enhanced the power-handling capabilities of semiconductor devices. A bipolar transistor exhibits two regions of operation that must be avoided:

- *Dissipation region.* Where the voltage-current product remains unchanged over any combination of voltage (V) and current (I). Gradually, as the collector-to-emitter voltage increases, the electric field through the base region causes hot spots to form. The carriers can actually punch a hole in the junction by melting silicon. The result is a dead (shorted) transistor.
- *Secondary breakdown* ($I_{s/b}$) *region.* Where the power transistor dissipation varies in a non-linear inverse relationship with the applied collector-to-emitter voltage when the transistor is forward-biased. The $I_{s/b}$ point on the power transistor safe operating area specification chart is the *inflection point* at which the secondary breakdown phenomenon occurs. Figure 4-10 shows a power transistor SOA curve.

To get SOA data into some type of useful format, a family of curves at various operating temperatures must be developed and plotted on a linear graph. This exercise gives a clear picture of what the data sheet indicates, compared to what happens in actual practice.

4.2.6 Power Handling Capability

The primary factor in determining the amount of power a given device can handle is the size of the active junctions on the chip. The

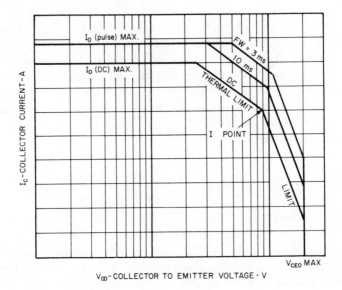

Figure 4-10 Safe operating area (SOA) curve for a typical power transistor.
(*Source: Broadcast Engineering Magazine.*)

same power output from a device may be achieved through the use of several smaller chips in parallel. This approach, however, can result in unequal currents and uneven distribution of heat. At high power levels, heat management becomes a significant factor in chip design.

Specialized layout geometries have been developed to ensure even current distribution throughout the device. One approach involves the use of a matrix of emitter resistances constructed so that the overall distribution of power among the parallel emitter elements results in even thermal dissipation. Figure 4-11 illustrates the *interdigited* geometry technique.

With improvements in semiconductor fabrication processes, output device SOA is primarily a function of the size of the silicon slab inside the package. Package type, of course, determines the ultimate dissipation because of thermal saturation with temperature rise. A good TO-3 or a 2-screw-mounted plastic package will dissipate approximately 350-375 W if properly mounted. Figure 4-12 demonstrates the relationships between case size and power dissipation for a TO-3 package.

4.3   Field-Effect Transistors

Field-effect transistors (FETs) are constructed with a conducting channel terminated by *source* and *drain* electrodes, and a *gate* termi-

RESISTORS

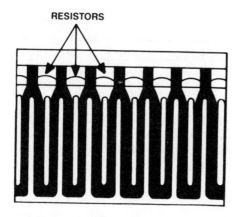

Figure 4-11 Interdigited geometry of emitter resistors used to balance currents throughout a power device chip.

nal in between. The conductive channel is effectively widened or narrowed by the electric field between the gate and each portion of the channel. No gate current is required for steady-state control.

Current flow in the channel is by majority carriers only, analogous to current flow in a resistor. The onset of conduction is not limited by diffusion speeds, only by the electric field accelerating the charged electrons. FETs typically use silicon as the semiconducting material.

FETs are made both in *p-channel* and *n-channel* configurations. An n-channel FET has a positive drain voltage with respect to source voltage; a positive increase in gate-to-source voltage increases drain current. Reverse polarities exist for p-channel devices.

The input impedance of a FET is very high and is primarily capacitive. The input capacitance consists of the gate-source capacitance in

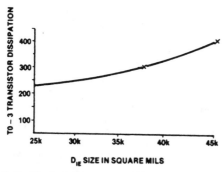

Figure 4-12 Relationship between case (die) size and power dissipation. (*Source: Broadcast Engineering Magazine.*)

parallel with the gate-drain capacitance multiplied by the gate gain + 1, assuming the FET has its source at an ac-ground potential. The high gate input impedance makes it possible to design broadband amplifiers with simple input matching networks. Also, the FET gate-source impedance remains capacitive to higher frequencies than bipolar devices, making internal matching networks unnecessary to VHF, even for devices of 100-150 W output power.

The *transconductance* (or gain) of a FET is defined as the ratio of the change in drain current to an accompanying small change in applied gate-to-source voltage. Other important parameters include:

- *On-resistance.* The ohmic value of the current channel from drain to source when the device is switched on. This is an important figure of merit because it determines the amount of current a device can handle without excessive dissipation.
- *Threshold voltage.* The lowest gate voltage at which a specified amount of drain current begins to flow. A current flow of 1 mA is commonly used as the measurement parameter.
- *Switching characteristics.* Device turn-on/off and rise/fall times. FET switching speeds are fast, relative to comparably sized bipolar transistors. Still, the physical structure of a FET results in parasitic capacitances between terminals which must be charged and discharged to effect the switching function. The impedance of the drive source has a significant effect on switching times. Table 4-2 lists the key parameters that describe FET operation.

FETs have two basic modes of operation: *depletion* and *enhancement*. Depletion refers to the decrease of carriers in the channel due to variation in gate voltage. The enhancement mode refers to the increase of carriers in the channel as a result of the application of a gate voltage. FET devices may also be constructed to operate in both the depletion and enhancement modes. A depletion-mode device will have drain current flow with zero gate voltage. Drain current is reduced by applying a reverse voltage to the gate. A depletion FET is not characterized with forward gate voltage.

4.3.1 Types of Devices

Two general types of FETs are commonly available: *junction FET* (JFET) and *insulated-gate FET* (IGFET) devices. The IGFET may be based on a variety of structures, including metal-insulator-semiconductor (MISFET) and metal-oxide-semiconductor (MOSFET).

Table 4-2 Basic field-effect transistor amplifier configurations and operating characteristics. (*Source:* Rohde and Bucher, *Communications Receivers: Principles and Design,* McGraw-Hill, New York, 1988.)

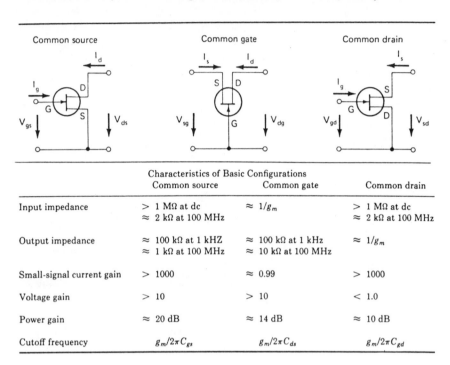

	Characteristics of Basic Configurations		
	Common source	Common gate	Common drain
Input impedance	> 1 MΩ at dc ≈ 2 kΩ at 100 MHz	$\approx 1/g_m$	> 1 MΩ at dc ≈ 2 kΩ at 100 MHz
Output impedance	≈ 100 kΩ at 1 kHZ ≈ 1 kΩ at 100 MHz	≈ 100 kΩ at 1 kHz ≈ 10 kΩ at 100 MHz	$\approx 1/g_m$
Small-signal current gain	> 1000	≈ 0.99	> 1000
Voltage gain	> 10	> 10	< 1.0
Power gain	≈ 20 dB	≈ 14 dB	≈ 10 dB
Cutoff frequency	$g_m/2\pi C_{gs}$	$g_m/2\pi C_{ds}$	$g_m/2\pi C_{gd}$

Figure 4-13 shows a cross section of a p-channel JFET. Channel current is controlled by reverse-biasing the gate-to-channel junction so that the depletion region reduces the effective channel width. The input impedance of the device is high because of the reverse-biased

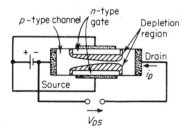

Figure 4-13 Construction of a p-channel junction field-effect transistor. (*Source:* Fink and Christiansen, *Electronics Engineers' Handbook, 3rd Ed.,* McGraw-Hill, New York, 1989.)

diode in the input circuit. In some respects, the volt-ampere characteristics are similar to a vacuum tube.

Figure 4-14 shows a cross section of a p-channel MOSFET. The device operates in the depletion mode. For zero gate voltage, there is no channel and the drain current is minimal. A negative voltage on the gate repels electrons from the surface and produces a p-type conduction region under the gate element. Compared with the JFET, the MOSFET has a wider gain-bandwidth product, and higher input impedance.

4.3.2   Circuit Configurations

As with bipolar transistors, three basic circuit configurations may be used with FETs:

- *Common-source*, where the source lead is tied to ground (the most common configuration). This arrangement is analogous to the common-emitter design.
- *Common-gate*, where the gate lead is tied to ground (analogous to the common-base).
- *Common-drain*, where the drain lead is tied to ground (analogous to the common-collector).

The output impedance of a common-source FET is primarily capacitive, as long as the drain voltage is above a *critical value*. For a junction-gate FET, this critical value is equal to the sum of *pinch-off* voltage (the potential that causes the device to stop conducting) and the gate-bias voltage. Actual FETs have a high drain resistance in parallel with this capacitance. At low drain voltages (near zero volts), the drain impedance of an ideal FET is a resistor, reciprocal in value to the transconductance of the FET in series with the residual end resistances between the source and drain terminals and the conducting FET channel. This permits a FET to be used as a vari-

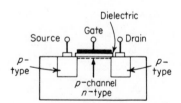

Figure 4-14 Construction of a p-channel MOS device. (*Source:* Fink and Christiansen, *Electronics Engineers' Handbook, 3rd Ed.,* McGraw-Hill, New York, 1989.)

able resistor in circuits controlling analog signals. At drain voltages between zero and the critical voltage, the drain current will increase with both increasing drain and gate voltage. This factor will cause increased saturation voltages in power amplifier circuits when compared to circuits with bipolar transistors.

Power Devices. Silicon RF power FETs are generally n-channel MOS enhancement mode devices. Most are *vertical structures,* meaning that the current flow is primarily vertical through the chip, with the bottom forming the drain contact. Vertical construction has the advantage of providing greater current density, which translates to more watts per unit area of silicon.

The design of a FET RF power amplifier has much in common with a bipolar amplifier. Both must include circuitry to supply bias voltages and matching networks to perform the necessary input/output impedance transformations over the operating frequency band. Most FET amplifiers produced today use the same basic collector voltages as bipolar systems (12.5 V, 28 V and 50 V). Higher voltage FET devices are also available. There is no FET parallel to the common zero base bias bipolar RF amplifier. FET amplifiers require forward gate bias for optimum power output and gain. The bias network may consist of a simple resistive divider.

MOSFET devices can usually be operated in parallel for higher output power at frequencies up to 150-200 MHz. Circuit instabilities can sometimes arise at higher frequencies, however, unless careful attention is given to amplifier design and component layout.

4.3.3   Operating Limits

Power MOSFETs have found numerous applications in RF transmission equipment because of their unique performance attributes. MOSFETs do not suffer from secondary breakdown, as bipolar transistors do. A variety of specifications can be used to indicate the maximum operating voltages a specific device can withstand. The most common specifications include:

- Gate-to-source breakdown voltage.
- Drain-to-gate breakdown voltage.
- Drain-to-source breakdown voltage (typically only used for MOSFET devices).

These limits mark the maximum voltage excursions possible with a given device before failure. Excessive voltages cause carriers within the depletion region of the reverse biased PN junction to acquire sufficient kinetic energy to cause ionization. Voltage breakdown can also occur when a critical electric field is reached. The magnitude of this voltage is determined primarily by the characteristics of the die itself.

Safe Operating Area. The safe dc operating area of a MOSFET is determined by the rated power dissipation of the device over the entire drain-to-source voltage range (up to the rated maximum voltage). The maximum drain-source voltage is a critical parameter. If exceeded even momentarily, the device can be damaged permanently. Figure 4-15 shows a representative SOA curve for a MOSFET. Notice that limits are plotted for several parameters, including drain-source voltage, thermal dissipation (a time-dependent function), package capability and drain-source on-resistance. The capability of the package to withstand high voltages is determined by the construction of the die itself, including bonding wire diameter, size of the bonding pad and internal thermal resistances. The drain-source on-resistance limit is simply a manifestation of Ohm's Law with a given on-resistance, current is limited by the applied voltage.

To a large extent, the thermal limitations described in the SOA chart determine the boundaries for MOSFET use in linear applications. The maximum permissible junction temperature also affects the pulsed current rating when the device is used as a switch.

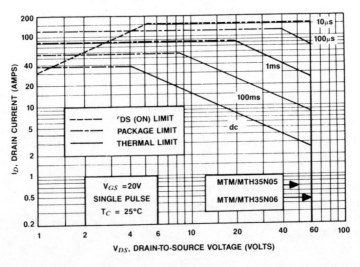

Figure 4-15 Safe operating area (SOA) curve for a power FET device.

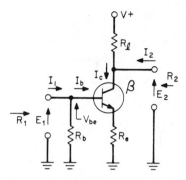

Figure 4-16 Common-emitter bipolar amplifier circuit. (*Source:* K. Blair Benson, *Audio Engineering Handbook,* McGraw-Hill, New York, 1988.)

MOSFETs are, in fact, more like rectifiers than bipolar transistors with respect to current ratings; their peak current ratings are not gain limited, but thermally limited.

In switching applications, total power dissipation is comprised of both switching losses and on-state losses. At low frequencies, switching losses are small. As the operating frequency increases, however, switching losses become significant factors in circuit design.

4.4 Analyzing Amplifier Characteristics

A solid state amplifier can best be described as using a single NPN transistor connected in a common-emitter configuration (see Figure 4-16). The *n*-channel FET version of this amplifier, connected in a common-source configuration, is shown in Figure 4-17.

4.4.1 DC Conditions

At zero frequency or dc (also at low frequencies), the transistor or FET amplifier stage requires an input voltage (E_1) equal to:

$E_1 = V_{be} + V_{Re}$ for a bipolar stage, or
$E_1 = V_{gs} + V_{Rs}$ for a FET stage

Where:
V_{be} = transistor base-emitter input voltage
V_{gs} = FET gate-source input voltage
V_{Re} = voltage across resistance R_e
V_{Rs} = voltage across resistance R_s

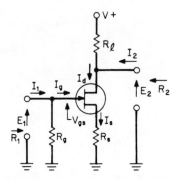

Figure 4-17 Common-source *n*-channel FET amplifier circuit. (*Source:* K. Blair Benson, *Audio Engineering Handbook,* McGraw-Hill, New York, 1988.)

Input current I_1 to the amplifier stage is equal to:

$I_1 = I_{Rb} + I_b$ for a bipolar stage, or
$I_1 = I_{Rg} + I_g$ for a FET stage

Where:
I_{Rb} = current through resistor R_b
I_{Rg} = current through resistor R_g
I_b = transistor base current
I_g = FET gate current

In most FET circuits the gate current is so small that it can be neglected.

In transistor circuits the base current (I_b) is equal to the collector current (I_c) divided by the current gain (beta) of the transistor. The input resistance (R_1) to the amplifier stage is equal to the ratio of input voltage (E_1) to input current (I_1). The input voltage and input resistance of a common-emitter or common-source amplifier stage increases as the value of the emitter or source resistor becomes larger. The output voltage (E_2) of the amplifier, operating without any external load, is equal to:

$E_2 = V - (R_L \times I_c)$ for a bipolar transistor, or
$E_2 = V - (R_L \times I_d)$ for a FET

Where:
V = supply voltage
R_L = collector or drain load resistor
I_c = collector current
I_d = drain current

An external load will cause the device to draw an additional current, I_2, which will increase the output current.

As long as the collector-to-emitter voltage is larger than the saturation voltage of the transistor, collector current will be nearly independent of supply voltage. Similarly, the drain current of a FET will be nearly independent of drain-to-source voltage as long as the potential is greater than an equivalent saturation voltage. This saturation voltage is approximately equal to the difference between the gate-to-source voltage and *pinch-off voltage*. The pinch-off voltage is the bias potential that causes nearly zero drain current. On some FET data sheets, the pinch-off voltage is referred to as the *threshold voltage*.

As the supply voltage is reduced, the collector or drain current will decline until it reaches zero.

The output resistance (R_2) of a transistor or FET amplifier is equal to:

$$R_2 = \frac{C_{R1} \times [R_e + (V_{ce}/I_c)]}{C_{R1} + [R_e + (V_{ce}/I_c)]} \text{ for a bipolar transistor}$$

$$R_2 = \frac{C_{R1} \times [R_s + (V_{ds}/I_d)]}{C_{R1} + [R_s + (V_{ds}/I_d)]} \text{ for a FET}$$

Where:
C_{Rl} = collector or drain load resistance
R_e = emitter resistor
R_s = source resistor
V_{ce} = collector-to-emitter voltage
V_{ds} = drain-to-source voltage
I_c = collector current
I_d = drain current

In real-world devices, an additional resistor (the relatively large output resistance of the device) is connected in parallel with the output resistance of the amplifier stage.

The collector current of a single-stage transistor amplifier is equal to the base current multiplied by the current gain of the transistor. Because the current gain may be specified as tightly as a two-to-one range at one value of collector current, or may have just a minimum value, knowledge of the input current is usually not sufficient to specify the output current of a transistor.

4.4.2   Operating Parameters

As previously stated, the input impedance is equal to the ratio of input voltage to input current, and the output impedance is the ratio of output voltage to output current. As the input current increases, the output current into the external output load resistor will increase by the current amplification factor of the stage. The output voltage will decrease because the increased current flows from the collector or drain voltage supply source into the collector or drain of the device (in a common-emitter or common-source configuration). Therefore, voltage amplification (V_a) is a negative number having a magnitude equal to:

$$V_a = \frac{\text{Output Voltage Change}}{\text{Input Voltage Change}} \times -1$$

The magnitude of voltage amplification may also be calculated from:

$$V_a = \text{Device Transconductance } (G_m) \times \text{Load Resistance}$$

The foregoing is true as long as the emitter or source resistor is zero, or the resistor is bypassed with a capacitor that effectively acts as a short circuit for ac signals.

In a bipolar transistor, the transconductance is approximately equal to:

$$G_m = \text{Emitter Current} \times 39$$

The value 39 is equivalent to the charge of a single electron divided by the product of Boltzmann's constant and absolute temperature in degrees Kelvin.

The power gain of a device is equal to the ratio of output power to input power, often expressed in decibels. Voltage gain or current gain may be stated in decibels, but must be so marked.

AC Gain. A resistor in series with the emitter or source causes *negative feedback*, which reduces the voltage gain of the single amplifier stage and raises its input impedance (see Figure 4-18). When resistor R_e is bypassed with a capacitor C_e, the amplification factor will be high at high frequencies and will be reduced by approximately 3 dB at the frequency where the impedance of the capacitor is equal to the emitter or source input impedance. This impedance is,

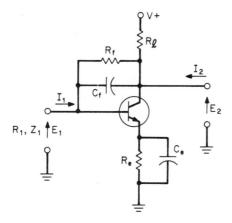

Figure 4-18 Single-stage bipolar amplifier with current and voltage feedback. (*Source:* K. Blair Benson, *Audio Engineering Handbook,* McGraw-Hill, New York, 1988.)

in turn, approximately equal to the inverse of the transconductance of the device. The gain of the stage will be approximately 3 dB higher than the dc gain at the frequency where the impedance of the capacitor is equal to the emitter or source resistor. This simplified analysis holds in cases where the product of transconductance and resistance values are much larger than 1.

A portion of the output voltage may also be fed back to the input (the base or gate terminal). This feedback resistor (R_f in Figure 4-18) will have the following effects on stage performance:

• Lower the input impedance
• Reduce current amplification
• Reduce output impedance
• Act as a supply voltage source for the base or gate

If the feedback element includes a capacitor (C_f in the previous diagram), high-frequency current amplification will be reduced by approximately 3 dB when the impedance of the capacitor is equal to the feedback resistor (R_f), and voltage gain of the stage is high. At still higher frequencies, amplification will decrease at the rate of 6 dB per octave. It should be noted that the base-collector or gate-drain capacitance of the device has the same effect of limiting high-frequency amplification. However, this capacitance becomes larger as the collector-base or drain-gate voltage decreases.

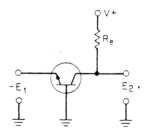

Figure 4-19 Common-base NPN amplifier circuit. (*Source:* K. Blair Benson, *Audio Engineering Handbook,* McGraw-Hill, New York, 1988.)

Feedback of a sample of the output voltage through an impedance lowers the input impedance of the amplifier stage. Voltage amplification will be affected only as this lowered input impedance loads the source of input voltage. If the source has a finite source impedance and the amplifier stage has very high voltage amplification and reversed phase, the effective amplification of this stage will approach the ratio of feedback impedance to source impedance and also have reversed phase.

Common-Base/Common-Gate Design. In a common-base bipolar transistor or common-gate FET stage, voltage amplification is the same as in the common-emitter or common-source connection (see Figure 4-19). However, the input impedance is approximately the inverse of the transconductance of the device. As a benefit, high-frequency performance will be improved because of the lower emitter-collector or source-drain capacitance, and the relatively low input impedance. The *cascade connection* of a common-emitter amplifier stage driving a common-base amplifier stage (as shown in Figure 4-20) exhibits nearly the dc amplification of a common-emitter with the wide bandwidth of a common-base stage. Common-base or common-gate amplifiers also offer stable operation at high frequencies (VHF and above), and provide for easy matching to RF transmission-line impedances, usually 50 or 75Ω.

Common-Collector/Common-Drain Design. The voltage gain of a transistor or FET is slightly below 1.0 for a common-collector or common-drain stage. The input impedance (Z_i) of the device is equal to:

$$Z_i = Z_l \times \beta + (1/\mu)$$

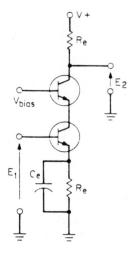

Figure 4-20 Cascade NPN amplifier circuit. (*Source:* K. Blair Benson, *Audio Engineering Handbook,* **McGraw-Hill, New York, 1988.)**

Where:
Z_l = load impedance
β = current gain
μ = transconductance of the device

Similarly, the output impedance of the stage (Z_o) is equal to:

$$Z_o = \frac{Z_i}{\beta + (1/\mu)}$$

Where:
Z_i = impedance of the source
β = current gain
μ = transconductance of the device

When resistors of identical value are connected between the collector and the supply voltage, and the emitter and ground, an increase in base voltage will result in an increase in emitter voltage that is nearly equal to the decrease in collector potential (for a bipolar transistor). Similarly for a FET stage, when resistors of identical value are connected between the drain and the supply voltage, and the source and ground, an increase in gate voltage will result in an increase in source voltage that is nearly equal to the decrease in drain

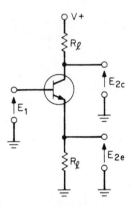

Figure 4-21 Split-load phase inverter circuit. (*Source:* K. Blair Benson, *Audio Engineering Handbook*, McGraw-Hill, New York, 1988.)

voltage. This type of connection is known as a *split-load phase inverter,* shown in Figure 4-21. Such a circuit is useful for driving push-pull amplifiers. It should be noted, however, that the impedances at the two output terminals are unequal.

The current gain of a transistor decreases at high frequencies as the emitter-base capacitance shunts a portion of the transconductance, thereby reducing gain until it reaches a value of 1 at the *transition frequency* of the device. As a result, the output impedance of an emitter-follower (common-collector) stage will increase with frequency. The net effect is an inductive source impedance when the input to the stage is resistive. If the source impedance is inductive, as it might be with cascaded-emitter followers, the output impedance can be a negative value at certain high frequencies. Amplifier oscillation may be a result. Similar considerations apply to common-drain FET stages.

Bias and Large Signals. The possibility for distortion products from an amplifier must be considered in any analog (non-switching) design. Although feedback can reduce distortion, it is necessary to ensure that each stage of amplification meets the following criteria:

• Normal input signals will not cause the amplifying device to operate with a (near) zero voltage drop across the component during any portion of the ac cycle (saturated state); and

• Normal input signals will not cause the amplifying device to operate with (near) zero current during any portion of the ac cycle (cut-off state).

Although described primarily with respect to a single-device-amplifier stage, the same holds true for any amplifier with multiple devices.

If a single-device-amplifier load consists of the collector or drain load resistor only, the operating point should be chosen so that in the absence of a signal, one-half of the voltage supply appears as the *quiescent voltage* across the load resistor.

Figure 4-22 illustrates the result if an additional resistive load (R_l') is connected to the output through a coupling capacitor (C_c). In such a case, the maximum peak load current (I_l) in one direction is equal to:

$$I_1 = \frac{V}{R_1 + R_1'} - I_q$$

Where:
V = collector supply voltage
I_q = collector quiescent current
R_l = collector resistance
$R_{l'}$ = load resistance

In the other direction, maximum load current is limited by the quiescent voltage across the device divided by the load resistance. The

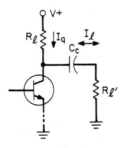

Figure 4-22 AC-coupled output circuit. (*Source:* K. Blair Benson, *Audio Engineering Handbook,* McGraw-Hill, New York, 1988.)

quiescent current flows in the absence of an ac signal and is the result of bias voltage or current only.

Because most signals have positive and negative peak excursions of equal probability, it is desirable for the two peak currents to be equal. This can be accomplished by increasing the quiescent current and/or changing the external resistances.

When two or more devices contribute current into an external load resistor, as shown in Figure 4-23, bias currents may be set so that the sum of all transconductances remains relatively constant. This provides for minimum stage distortion.

4.5 Reliability of Solid State Devices

A device can fail in a catastrophic, intermittent, or degraded mode. Such failures are usually opens, shorts, or parameters out of specifications. For semiconductors, the three most destructive stresses are excessive temperature, voltage, and vibration.

4.5.1 Semiconductor Failure Modes

Active components are the heart of any RF system, and most — with the exception of high power transmitter stages — employ semiconductors exclusively.

Failure modes can be broken down into two basic categories: mechanical (including temperature and vibration) and electrical (including electrostatic discharge and transient overvoltage). Semiconductor manufacturers are able to increase device reliability by analyzing why good parts go bad. There are, in fact, a frightening number of mechanical construction anomalies that can result in degraded or catastrophic failure of a semiconductor device. Some of the more significant threats include:

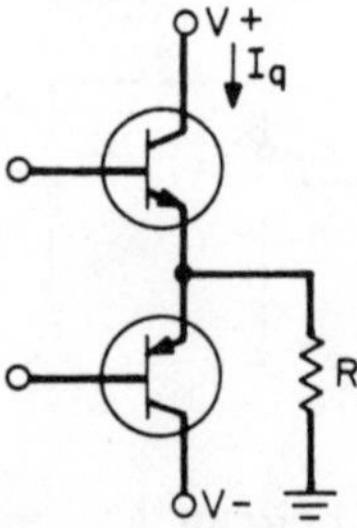

Figure 4-23 Push-pull half-bridge output circuit. (*Source:* K. Blair Benson, *Audio Engineering Handbook,* McGraw-Hill, New York, 1988.)

- Encapsulation failures caused by humidity and impurity penetration, imperfections in termination materials, stress cracks in the encapsulation material, and differential thermal expansion coefficients of the encapsulant, device leads, or chip.
- Wire bond failures caused by misplaced bonds, crossed wires, and oversize bonds.
- Imperfect chip attachment to the device substrate resulting in incomplete thermal contact, stress cracks in the chip or substrate, and solder or epoxy material short circuits.
- Aluminum conductor faults caused by metalization failures at contact windows, electromigration, corrosion, and geometric misalignment of leads and/or the chip itself.

The principal failure modes for semiconductor devices include the following:

- Internal short circuit between metalized leads or across a junction, usually resulting in system failure.
- Open circuit in the metalization or wire bond, usually resulting in system failure.
- Variation in gain or other electrical parameters, resulting in marginal performance of the system or temperature sensitivity.
- Leakage currents across P-N junctions, causing effects ranging from system malfunctions to out-of-tolerance conditions.
- Shift in turn-on voltage, resulting in random logic malfunctions in digital systems.
- Loss of seal integrity through the ingress of ambient air, moisture, and/or contaminants. The effects range from system performance degradation to complete failure.

4.5.2   Stress Induced Failures

Thermal fatigue represents a significant threat to any semiconductor component, especially power devices. This phenomenon results from the thermal mismatch between a silicon chip and the device header under temperature-related stresses. Typical failure modes include voids and cracks in solder material within the device, which results in increased thermal resistance to the outside world and the formation of *hot spots* inside the device. Catastrophic failure will occur if sufficient stress is put on the die. Figure 4-24 illustrates the defor-

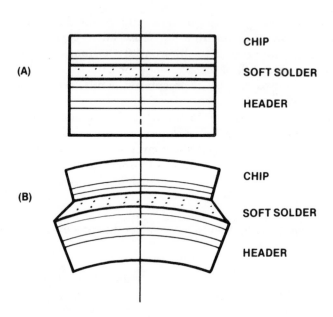

Figure 4-24 The mechanics of thermal stress on a semiconductor device: (a) normal chip/solder/header composite structure; (b) the composite structure subjected to a change in temperature. (*Source: Broadcast Engineering Magazine.*)

mation process that results from excessive heat on a semiconductor device.

As a case in point, consider the 2N3055 bipolar transistor. The component is an NPN power device rated for 115 W dissipation at 25°C ambient temperature. The component can handle up to 100 V on the collector at 15 A. Although the 2N3055 is designed for demanding applications, the effects of thermal cycling take their toll. Figure 4-25 charts the number of predicted thermal cycles for the 2N3055-vs.-temperature change. Note that the lifetime of the device increases from 9,000 cycles at 120°C to 30,000 cycles at 50°C.

4.5.3 Breakdown Modes

It is estimated that as much as 95 percent of all transistor failures are directly or indirectly the result of excessive dissipation or applied voltages in excess of maximum design limits. There are at least four types of voltage breakdown that must be considered in a reliability analysis of discrete power transistors. Although each type is not

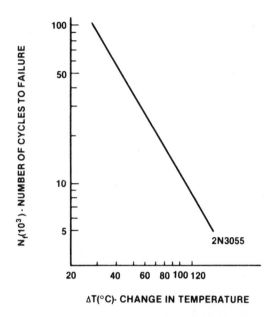

Figure 4-25 The effects of thermal cycling on a 2N3055 power transistor. (*Source: Broadcast Engineering Magazine.*)

strictly independent, they can be treated separately, keeping in mind that each is related to the others.

Avalanche Breakdown. A voltage breakdown that occurs in the collector-base junction, similar to the *Townsend effect* in gas tubes. This effect is caused by the high dielectric field strength that occurs across the collector-base junction as the collector voltage is increased. This high intensity field accelerates the free charge carriers so they collide with other atoms, knocking loose additional free charge carriers that, in turn, are accelerated and have further collisions. This multiplication process occurs at an increasing rate as the collector voltage increases until at some voltage, V_a (avalanche voltage), the current suddenly tries to go to infinity. If enough heat is generated in this process, the junction will be damaged or destroyed. A damaged junction will result in higher than normal leakage currents, increasing the steady-state heat generation of the device, which may ultimately destroy the semiconductor junction.

Alpha Multiplication. A type of breakdown closely related to the avalanche effect. Alpha multiplication is produced by the same physical phenomenon that produces avalanche, but differs in circuit configuration. This effect occurs at a lower potential than the avalanche breakdown voltage and generally is responsible for collector-emitter breakdown when base current is zero.

Punch-Through. A voltage breakdown occurring between the collector-base junction because of high collector voltage. As collector voltage is increased, the *space charge region* (collector junction width) gradually increases until it penetrates completely through the base region, touching the emitter. At this point the emitter and collector are effectively shorted together. This type of breakdown occurs in some PNP junction transistors. Generally, however, alpha multiplication breakdown occurs at a lower voltage than punch-through. Because this breakdown occurs between collector and emitter, punch-through is more serious in the common-emitter or common-collector configuration.

Thermal Runaway. A regenerative process where an increase in temperature causes an increase in leakage current that results in increased collector current, which, in turn, causes greater power dissipation. This action raises the junction temperature, causing a further increase in leakage current. If the leakage current is high enough (resulting from high temperature or high voltage), and the current is not adequately stabilized to counteract increased collector current because of increased leakage current, the process can regenerate to a point that the temperature of the transistor (and power dissipation) rapidly increases, destroying the device. This failure mode is more prominent in high power transistors, where the junction is normally operated at elevated temperatures and where high leakage currents are present because of the large junction area. Thermal runaway is related to the avalanche effect, and is dependent upon circuit stability, ambient temperature, and transistor power dissipation.

Effects of Breakdown. The effects of the breakdown modes outlined manifest themselves in various ways on power semiconductor devices. Avalanche breakdown usually results in the destruction of the collector-base junction because of excessive currents, which, in turn, results in an open circuit between collector and base.

Breakdown due to alpha multiplication and thermal runaway most often results in destruction of the transistor because of excessive heat dissipation. The outward failure mode is usually a short between the collector and emitter. This condition, which is most common in transistors that have suffered catastrophic failure, is not always easily detected. In many cases an ohmmeter check may indicate a good device. Only after operating voltages are applied will the failure mode be exhibited.

Punch-through breakdown generally does not permanently damage a transistor. It can be a self-healing type of breakdown. After the over-voltage is removed, the transistor may operate satisfactorily.

4.5.4 Heat Management

Heat generated in a power semiconductor must be removed at a sufficient rate to keep the junction temperature within a specific upper limit. This is accomplished primarily by conduction from the junction through the transistor material to a metal mounting base, that is designed to provide good thermal contact, to an external heat dissipator or heat sink.

Because heat transfer is associated with a temperature difference, a differential will exist between the collector junction and the transistor mounting surface. A temperature differential will also exist between the device mounting surface and the heat sink. Ideally these differentials will be small. They will, however, exist to one extent or another. It follows, therefore, that an increase in dissipated power at the collector junction will result in a corresponding increase in junction temperature. In general, assessing the heat sink requirements of a device or system (and the potential for problems) is a difficult proposition.

Figure 4-26 shows some of the primary elements involved in thermal transmission of energy from a silicon junction to an external heat sink. An electrical analog of the process is helpful for illustration. The model shown in Figure 4-27 includes two primary elements: *thermal capacitance* and *thermal resistance*. The energy storage property of a given mass, expressed as C, is the basis for the transient thermal properties of transistors. The thermal transmission loss from one surface or material to another, expressed as θ, causes a temperature differential between the various components of the semiconductor model shown in the figure.

Although this model may be used to predict the rise of junction temperature that results from a given increase in power dissipation, it is an extreme over-simplification of the mechanics involved. The elements considered in our example include the silicon transistor die (Si); the solder used inside the device to bond the emitter, base, and collector to the outside-world terminals; and the combined effects of the heat sink and transistor case. This model assumes the transistor

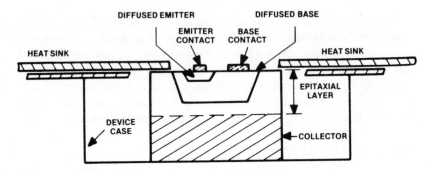

Figure 4-26 Simplified model of thermal transmission from the junction of a power transistor (TO-3 case) to a heat sink. (*Source: Broadcast Engineering Magazine.*)

is directly mounted onto a heat sink, not through a mica (or other style) insulator. A similar model can be developed for FET devices.

The primary purpose of a heat sink is to increase the effective heat-dissipation area of the power semiconductor. If the full power-handling capability is to be achieved, there must be a zero temperature differential between the case and the ambient air. This condition exists only when the thermal resistance of the heat sink is zero,

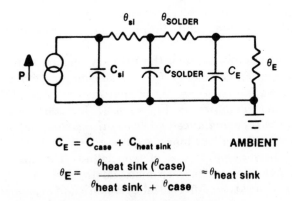

$$C_E = C_{case} + C_{heat\ sink}$$

$$\theta_E = \frac{\theta_{heat\ sink}\ (\theta_{case})}{\theta_{heat\ sink}\ +\ \theta_{case}} \approx \theta_{heat\ sink}$$

P = THERMAL ENERGY GENERATED BY DEVICE

θ = THERMAL RESISTANCE (MEASURED IN °C/WATT)

C = THERMAL CAPACITANCE (MEASURED IN WATT-SECONDS/°C)

Figure 4-27 Simplified electrical equivalent of the heat transfer mechanism from the junction of a power transistor to an external heat sink. (*Source: Broadcast Engineering Magazine.*)

requiring an infinitely large heat sink. Although such a device can never be realized, the closer the approximation of actual conditions to ideal conditions, the greater the maximum possible operating power.

In typical power transistor applications, the case must be electrically insulated from the heat sink (except for circuits where the collector or drain is grounded). The thermal resistance from case to heat sink, therefore, includes two components: surface irregularities of the insulating material, transistor case, and heat sink; and the insulator itself. Thermal resistance resulting from surface irregularities can be minimized through the use of silicon grease compounds. Thermal resistance of the insulator itself, however, can represent a significant problem. Unfortunately, materials that are good electrical insulators are also usually good thermal insulators. The best materials for such applications are mica, beryllium oxide, and anodized aluminum.

4.5.5   Transient Overvoltages

Semiconductor failures caused by high voltage stresses are becoming a serious problem for system users, as sensitive devices find their way into transmission circuits. Failures are the result of two primary overvoltage sources:

- *External man-made and natural.* Man-made overvoltages are commonly the result of ac utility company switching and system faults. Natural sources include lightning and other atmospheric disturbances.
- *Electrostatic discharge* (ESD). Depending upon the local weather conditions, damaging ESD potentials can be developed by simply walking across a carpeted floor and then touching a component.

Transient Disturbances. Short-term ac-voltage disturbances may be divided into four basic categories, as shown in Figure 4-28. The generally accepted definitions for these disturbances are:

- *Voltage surge.* An increase of 10-35 percent above the normal line voltage for a period of 16 ms to 30 s.
- *Voltage sag.* A decrease of 10-35 percent below the normal line voltage for a period of 16 ms to 30 s.
- *Transient disturbance.* A voltage pulse of high energy and short duration impressed upon the ac waveform. The overvoltage pulse

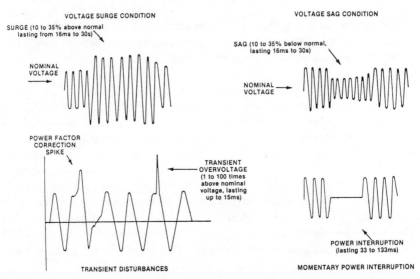

Figure 4-28 The four basic classifications of short-term power-line disturbances. (*Source: Broadcast Engineering Magazine.*)

may be one to 100 times the normal ac potential and may last up to 15 ms. Rise times can measure in the nanosecond range.

- *Momentary power interruption.* A decrease to zero voltage of the ac power line potential, lasting from 33-133 ms. (Longer-duration interruptions are considered power outages.)

Of these ac line disturbances, transients are by far the most damaging to transmission equipment. Transmitters are especially vulnerable because they are invariably connected to an antenna of some kind, which is exposed to lightning discharges. Figure 4-29 illustrates the susceptibility of semiconductors to transient discharges.

Failure Modes. Semiconductor devices can be destroyed or damaged by transient disturbances in one of several ways. A high reverse voltage applied to a nonconducting PN junction can cause avalanche currents to flow, heating the junction irregularly and consequently releasing more carriers, which conduct added current in the heated junction area. If enough heat is generated in this process, the junction can be damaged or destroyed. If such a process occurs between the base and emitter junctions of a transistor, the effect may be either minor or catastrophic. With a minor failure, the gain of the transistor can be reduced by the creation of *trapping centers,* which

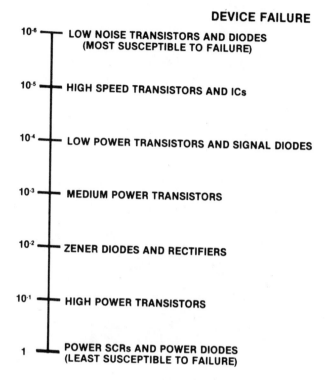

JOULES

DEVICE FAILURE

- 10^{-6} — LOW NOISE TRANSISTORS AND DIODES (MOST SUSCEPTIBLE TO FAILURE)
- 10^{-5} — HIGH SPEED TRANSISTORS AND ICs
- 10^{-4} — LOW POWER TRANSISTORS AND SIGNAL DIODES
- 10^{-3} — MEDIUM POWER TRANSISTORS
- 10^{-2} — ZENER DIODES AND RECTIFIERS
- 10^{-1} — HIGH POWER TRANSISTORS
- 1 — POWER SCRs AND POWER DIODES (LEAST SUSCEPTIBLE TO FAILURE)

Figure 4-29 An estimate of the susceptibility of semiconductor devices to failure because of transient energy. This estimate assumes a transient duration of several microseconds. (*Source: Broadcast Engineering Magazine.*)

restrict the free flow of carriers. These trapping centers are created by avalanche-damaged emitter-base junctions. With a catastrophic failure, the transistor will cease to function altogether.

Thermal runaway triggered by a sudden increase in gain is another transient-caused failure possible in a semiconductor device. The increased gain can result from the heating effect of a transient on a transistor, which may bring the device (operating in the active region) out of its safe operating area.

The oscillating and decaying tail of many transient disturbances can also subject semiconductor devices to severe voltage polarity reversals, forcing the components into or out of a conducting state. This action can damage the semiconductor junction or result in catastrophic failure of the component.

Electrostatic Discharge. Component failures caused by ESD are limited primarily to low power FETs and integrated circuits. It should be noted, however, that an ESD of sufficient potential can also damage a high power RF transistor or MOSFET. Figure 4-30 shows an electron microscope photo of a semiconductor device that failed because of out-of-tolerance electrical conditions. An electrostatic discharge to this MOSFET damaged the metalization connection point of the device, resulting in catastrophic failure. Note the

Figure 4-30 Scanning electron microscope photo illustrating ESD damage to the metalization of a metal-oxide semiconductor FET. (*Source: Broadcast Engineering Magazine.*)

places where damage occurred. The objects in the photo that look like bent nails are actually gold lead wires.

The most common cause of failure in a power MOSFET is an overvoltage that exceeds the maximum rated drain-source voltage of the device. Load transients caused by switching high currents or from lightning discharges may contain enough energy to destroy a device if it begins to avalanche.

Bibliography

1. Weirather, Robert, "Power Semiconductors in Today's Transmitters," *Proceedings of the 1989 SBE National Convention*, Kansas City, October 1989.
2. Fink, D. and D. Christiansen, *Electronics Engineer's Handbook*, Third Edition, McGraw-Hill, New York, 1989.
3. Jordan, Edward C., *Reference Data for Engineers: Radio, Electronics, Computer and Communications*, Seventh Edition, Howard W. Sams Company, Indianapolis, IN, 1985.
4. RF Transistor Design (an application note), *RF Device Data, Volume II*, Motorola, Phoenix, AZ, 1988.
5. Application note AN211A, Field Effect Transistors in Theory and Practice, *RF Device Data, Volume II*, Motorola, Phoenix, AZ, 1987.
6. Application note AN878, VHF MOS Power Applications, *RF Device Data, Volume II*, Motorola, Phoenix, AZ, 1988.
7. Granberg, Helge, application note AR165S, RF Power MOSFETs, *RF Device Data, Volume II*, Motorola, Phoenix, AZ, 1988.
8. *Power MOSFET Transistor Data Handbook*, Motorola, Phoenix, AZ, 1988.

5

Applying Semiconductor RF Devices

5.1 Introduction

There are any number of ways to construct a solid state amplifier for RF applications. The primary criteria for design include:

- Output power
- Operating frequency
- Required bandwidth
- Noise performance
- Efficiency desired
- Power supply voltages available
- Cost

Not surprisingly, achieving all of the design objectives is a difficult task. Trade-offs are usually required to produce a practical system. The most basic design variable is the type of power device used. For solid state RF applications the primary choices are bipolar transistors and FETs. Table 5-1 lists some of the principle comparison parameters.

5.1.1 Circuit Types

The common-emitter (or common-source FET) configuration is the building block of most solid state systems. In many circuits, combi-

Table 5-1 Comparison of bipolar and FET operating parameters.

Parameter	Bipolar	FET
Current carrier	Minority	Majority
Input impedance	Medium	High
Switching speed	Medium	High
Ruggedness	Good	Excellent
Gain	Medium/low	High
Paralleling capability	Difficult	Easy
Thermal handling capability	Excellent	Fair

nations of all three basic topologies (common-emitter/source, common-base/gate, and common-collector/drain) can be found. Circuit types can be classified into two basic categories: *symmetrical* and *asymmetrical*. In a symmetrical scheme, both the positive and negative signal swings are driven equally. This ensures that the inter-element capacitances of the output stage and frequency compensation networks do not unbalance the signal, causing harmonic distortion. When the output has equal drive, the large-signal frequency response of the positive and negative halves should also be equal. For this reason, the amplifier should exhibit a symmetrical *slew rate.* (Slew rate is a measure of how fast a signal changes from one instantaneous value to another.) These are important criteria for linear amplification. An asymmetrical drive scheme is one in which the positive and negative signal swings are not equal. Active current source loads and boot-strapped resistive collector loads, unless carefully designed, will not deliver the same low-distortion performance as a symmetrical drive scheme. Low distortion is important in systems requiring linear amplification. In pulsed systems (including class C operation), low distortion usually takes a back seat to efficiency. Harmonic distortion products are removed by RF filters after the amplifying device.

Design Examples. There are nearly as many circuit topologies for power amplification as there are companies producing amplifiers. The most popular types of output stage designs fall into one of the following categories:

- *Single-ended common-emitter/common-source.* This is the simplest of all amplifier designs. Operating modes include class A for linear RF systems and class C for pulsed or FM systems.

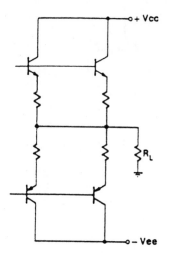

Figure 5-1 Power supply ground-referenced conventional, parallel-connected unity gain output stage. (*Source: Broadcast Engineering Magazine.*)

- *Parallel device amplification.* A parallel device circuit is essentially a tandem version of the single-ended amplifier.
- *Push-pull.* The basic class B or AB dual device linear amplifier.
- *Single- or multi-device switching.* The single-device switching configuration is the basic high efficiency pulsed topology, typically operating in a class D mode. This approach may be used with a proper RF tank and filtering for AM, FM, and pulsed systems. Multiple devices are used to achieve higher power output levels.

Amplifier types may be further categorized by the type of ground and power supply references. Common configurations include:

- *Conventional chassis referenced.* A parallel-connected unity-gain stage is show in Figure 5-1. A series-parallel-connected unity gain circuit is shown in Figure 5-2. An output stage with gain is shown in Figure 5-3. A high efficiency class G output circuit with unity gain is shown in Figure 5-4.
- *Floating bridge output with no chassis ground reference* (see Figure 5-5).
- *Bridge output with chassis ground reference.* This approach necessitates a floating power supply, as shown in Figure 5-6.

With many of these approaches, the actual circuit can be implemented with series-connected output devices, parallel-connected out-

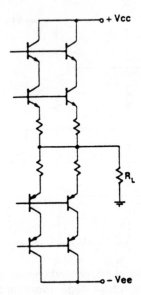

Figure 5-2 Power supply ground-referenced series-parallel-connected output stage with unity gain. (*Source: Broadcast Engineering Magazine.*)

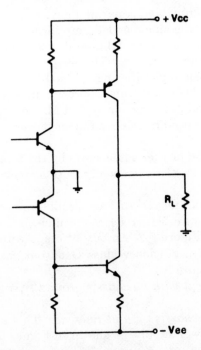

Figure 5-3 Power supply ground-referenced output stage with gain. (*Source: Broadcast Engineering Magazine.*)

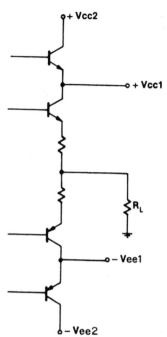

Figure 5-4 Power supply ground-referenced high-efficiency class G output stage with unity gain. (*Source: Broadcast Engineering Magazine.*)

put devices, or a combination of series/parallel configurations. Usually the more esoteric approaches are used to design around some inherent component limitation, either real or self-imposed.

In some specialized cases, the conventional approach simply will not work. A high-voltage amplifier is a good case in point. The breakdown voltage limitations of most devices necessitate some type of series-connected output stage.

Circuit Operation. To understand how a stage functions it is necessary to first know the electrical conditions under which the power devices will operate. Figure 5-7 shows a simplified 2-transistor complementary output stage driving a resistive load. It can be shown mathematically that class B efficiency is not a linear function of output power. In a class B amplifier, maximum dissipation occurs at 40.5 percent of full output. Figure 5-8 shows the relationship of output current, load voltage, voltage across the transistors, and the power pulses (which have a frequency twice that of the output voltage frequency).

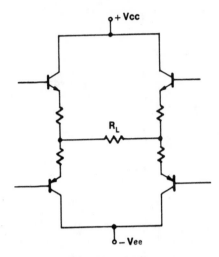

Figure 5-5 Floating bridge output stage with unity gain. (*Source: Broadcast Engineering Magazine.*)

Various classes of amplifiers have been developed to reduce output stage dissipation for both linear and pulsed operation. The class G amplifier (shown in Figure 5-9) consists of a series-connected output stage with diode switching to two different power supply levels. Another type of amplifier, class H (shown in Figure 5-10), uses switched power supplies in conjunction with a conventional output stage. These designs realize an approximate four-to-one reduction in maxi-

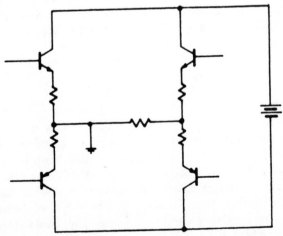

Figure 5-6 Floating power supply, ground-referenced bridge with unity gain. (*Source: Broadcast Engineering Magazine.*)

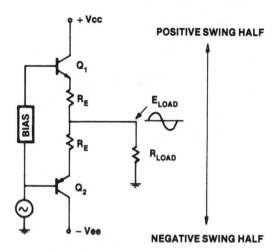

Figure 5-7 Simplified bipolar complementary output stage. (*Source: Broadcast Engineering Magazine.*)

mum worst-case dissipation. At full power, the two classes can be a few percentage points more efficient than class B. However, actual circuit implementation may limit the full-power dissipation to a value close to class B because of the added saturation losses of the additional series transistors.

The preceding examples assume the amplifier is feeding a resistive load. With a reactive load, the voltage/current relationships are dis-

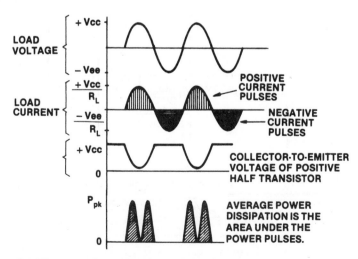

Figure 5-8 Time relationship of output stage current, voltage and dissipation. (*Source: Broadcast Engineering Magazine.*)

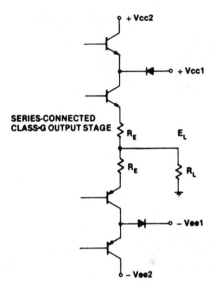

Figure 5-9 Typical class G power output circuit. (*Source: Broadcast Engineering Magazine.*)

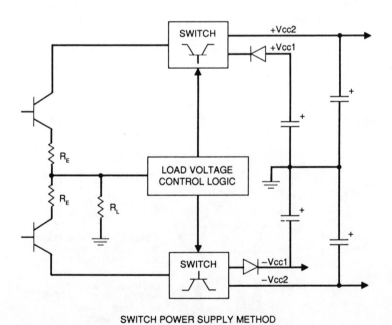

SWITCH POWER SUPPLY METHOD

Figure 5-10 Simplified class H power amplifier circuit. (*Source: Broadcast Engineering Magazine.*)

placed in time (out of phase) from the resistive case. The worst-case dissipation occurs with a 90° reactive load and can result in a 214 percent increase in output stage dissipation. Dissipation under this worst-case scenario increases from 40.5 percent of output power to 127 percent output power. In other words, more power is generated in heat than is delivered to the load.

Amplifier Load. The ideal load for almost any power amplifier is a resistive load that exhibits the following properties:

• All the power generated by the amplifier is delivered to the load without reflections back to the amplifier
• The load resistance does not change with frequency
• The load resistance does not change with power level

Most RF amplifiers do not operate into a perfect load. Antenna systems may present the power amplifier with a load that is a combination of resistive, inductive, and capacitive components.

Switching and Inductive-Load Ratings. When transistors are used to drive an inductive or resonant load, the current in the load inductor will tend to flow in the same direction, even if interrupted by the transistor. The resulting voltage spike caused by the collapse of the magnetic field may destroy the transistor unless protective measures are taken. Device failures can be prevented by installing protective diodes to shunt the spike current, or by installing a switching device that is overrated for the application.

Transistors are often used to switch high currents. The various junction capacitances are voltage-dependent in the same manner as the capacitance of tuning diodes (varactors), which have maximum capacitance at forward voltages, less capacitance at zero voltage, and lowest at reverse voltages. These capacitances, stray inductances, and circuit resistances may combine to increase switching turn-on/turn-off delay time. The type of device used and the frequency of operation will determine whether the delay time will be detrimental to proper circuit operation.

5.2 Solid State RF Amplifiers

Solid-state transmitters make use of various schemes employing pulse-type modulation and high-efficiency linear techniques. In order to achieve the needed RF output level, several semiconductor devices are usually assembled together in a single module. These 50 Ω building blocks are then combined to provide whatever power level is needed.

Because of the unpredictable conditions that some transmission equipment will experience over its operating lifetime, protection circuits are required. Each module must be protected against voltage transients on the V+ input line and RF output port, thermal overloads, and severe load VSWR. Some designs also incorporate a soft-start feature. With this approach, when the system is turned on, the modules do not simultaneously draw current from the power supply. Rather, each module switches on at a different time, thereby reducing the initial surge applied to the supply. The delay times may range from a few milliseconds to a few seconds.

Each module in a typical high power system is combined through a combining network. In addition to coupling the modules together to obtain higher power levels, the network also protects the modules from the load and from each other. In a properly designed combiner, the failure of one module will have no effect on the other operating modules.

5.2.1 Device Requirements

High power RF applications demand a high degree of performance from transistors and FETs. Operation at high frequencies requires a device geometry and package that offers the lowest possible stray inductance and capacitance. Operation at high power levels requires a device that can tolerate high temperatures and a package that offers low thermal resistance to its accompanying heat sink. Figure 5-11 shows some of the more common RF transistor packages. A variety of approaches have been taken to solve the problem of generating high power at high frequencies. A representative device is shown in Figure 5-12. This BLV37 transistor (Philips) is rated for 250 W dissipation. The package is a 5-lead rectangular flange envelope with a ceramic cap. All leads are isolated from the flange to facilitate a good thermal bond with the external heat sink. The transistor package (designated SOT-179) measures 13.2 x 28 mm. It includes a thin base layer of beryllium oxide between the transistor junction die and the package, which incorporates an *Alkonite* flange. Alkonite is a strong heat-conductive substance that can be machined flat for optimum contact with the heat sink. Internal input matching provides for wideband operation and high power gain.

Key specifications for the BLV37 transistor include:

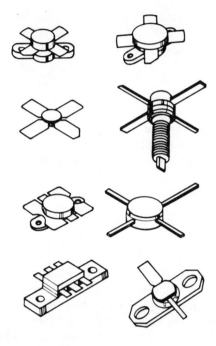

Figure 5-11 Common RF transistor packages.

Figure 5-12 Philips BLV37 RF power transistor. *(Courtesy of Philips.)*

Maximum power dissipation	250 W
Typical operating frequency	108 MHz
Nominal collector voltage	28 V
Minimum gain	11 dB
Typical operating class	Class B
Typical operating efficiency	65%

This device is intended for use in FM broadcast transmitters and commercial/military mobile radios. The BLV37 package contains two transistors in a push-pull configuration. Each device has 1,400 emitters on an area of just 2.5 × 4.8 mm. This arrangement provides for even current distribution on the die, thereby eliminating hot spots. Emitter ballasting resistors, over 700 per device, further improve heat distribution. An example RF amplifier using the BLV37 transistor is shown in Figure 5-13. The mechanical layout of the amplifier is detailed in Figure 5-14.

At UHF frequencies and above, transistor designers find increased difficulty in fabricating high power devices. As a general rule of thumb, as frequency goes up, power goes down. The PXB1650U (Philips) is a representative microwave power device. Rated for 50 W CW operation at 1.6 GHz, the transistor offers a collector efficiency as high as 52 percent and a typical power gain of 9.5 dB. The NPN device is designed to operate as a common-base class C narrowband

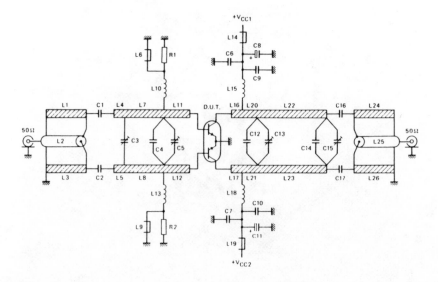

Figure 5-13 Example RF amplifier using the BLV37 transistor. This circuit will develop 250 W at 108 MHz. *(Courtesy of Philips.)*

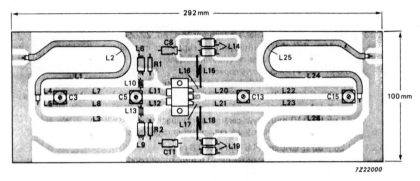

7Z22000

Figure 5-14 Physical layout of the 250 W amplifier based on the BLV37 transistor. (Courtesy of Philips.)

amplifier. Multiple emitters and diffused emitter ballasting resistors prevent the formation of hot spots on the die. The PXB1650C is intended for use in voice and data communications applications in ship, aircraft, and Earth station facilities.

Power MOSFET devices are finding increased applications in high power RF equipment. The MRF-156 MOSFET (Motorola) is a good example. Intended for use in military and commercial HF/SSB fixed, mobile, and marine transmitters, operating parameters for the MRF-156 include:

Maximum power output	600 W
Typical operating frequency	150 MHz
Nominal drain voltage	50 V
Minimum gain (measured at 30 MHz)	20 dB

Compared with bipolar transistors, RF power FETs typically exhibit greater gain, higher input impedance, enhanced thermal stability, and lower noise.

5.2.2   Stage Coupling

The typical gain of a single RF amplifier stage ranges from 5 to 12 dB, depending on the operating frequency and bandwidth. To obtain higher gain, stages are combined in series. Usually the input and output of each stage is designed to interface using a common impedance (typically 50 Ω resistive) to facilitate stage-to-stage compatibility and isolation. Band-limiting is usually a useful byproduct of interstage coupling.

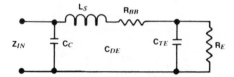

WHERE:
R_F = EMITTER DIFFUSION RESISTANCE
$C_{DE} \cdot C_{TF}$ = DIFFUSION AND TRANSITION
CAPACITANCES OF THE EMITTER JUNCTION
R_{BB} = BASE SPREADING RESISTANCE
C_C = PACKAGE CAPACITANCE
L_S = BASE LEAD INDUCTANCE

Figure 5-15 Equivalent circuit for the input impedance of an RF power transistor. *(Courtesy of Motorola.)*

Coupling networks are needed to provide for the best possible energy transfer from stage to stage. The input impedance of an RF power transistor is low, decreasing as the power increases, or as the chip size becomes larger. Impedance transformations ratios of 10-20 are not uncommon. Coupling circuits must deal with a number of parameters, not the least of which is operating bandwidth. Figure 5-15 shows the equivalent circuit for the input of a power transistor. Most VHF high-band transistors will have the series resonant frequency within their operating range. That is, the input will be purely resistive at one single frequency. The parallel resonant frequency will typically be outside the operating range of the device. The output impedance of an RF power transistor consists principally of a capacitance that varies as a function of operating frequency (Figure 5-16). The internal resistance of the device is generally much higher than the load, and is typically neglected. Strictly speaking, impedance matching is accomplished only at the input of the device. Interstage and load matching are basically impedance *transformations* of the device input impedance and of the load into a value determined by the power demanded from the output device and the supply voltage.

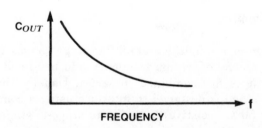

Figure 5-16 Transistor output capacitance as a function of frequency. *(Courtesy of Motorola.)*

Two techniques are commonly used to accomplish the required coupling: transformers and transmission line techniques, including *stripline*. For frequencies below about 100 MHz, matching is usually accomplished with ferrite transformers. Above 100 MHz, transformers are too inefficient to compete with simpler transmission line devices. Transmission line networks are only practical at VHF frequencies and above. Lower frequencies translate to longer wavelengths, and the length of coax or stripline required to yield a 1/2- or 1/4-wave section can become prohibitive.

Broadband Transformer. Impedance matching devices for RF applications can take on a number of forms. Most involve a specified number of primary and secondary turns around a given core material. Depending on the application, multiple secondary (or primary) windings may be used.

Figure 5-17 illustrates a common design intended for high transformation ratios (16 and greater). The low impedance winding consists of one turn, which limits the available ratios to integers 1, 4, 9 and so on. In the 16:1 design shown in the figure, two copper tubes form the primary winding. They are shorted on one end by a piece of copper-clad laminate with holes through which the tubes pass. A similar piece of laminate is soldered to the opposite ends of the tubes. The copper foil is divided into two sections, thus isolating the ends where the primary connections are made. The secondary winding is formed by threading wire through the tubes for the required number of turns.

Broadband transformers may also be used to form hybrid combiners, which sum the outputs of several power amplifiers to feed a single load. When physically reversed, most hybrids can also be used

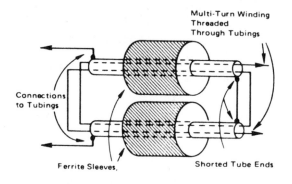

Figure 5-17 Physical construction of a 16:1 broadband transformer. *(Courtesy of Motorola.)*

as signal splitters. This concept is illustrated in Figure 5-18. Hybrids are intended to both combine (or distribute) power and to provide isolation between modules. Fault-tolerant designs require that in the event a module fails, the remaining modules in the system continue operating without degradation. Isolation is especially important when combining the outputs of linear amplifier modules because a constant load impedance must be maintained.

Transmission Line Matching. A fundamental law of transmission line theory states that if a transmission line is terminated in its characteristic impedance (usually 50 Ω in RF systems), the input impedance of the line will also be 50 Ω, regardless of the length of the cable. It is this property of a transmission line that makes it useful in transferring power to distant loads. An ideal match, however, is seldom achieved in practice.

There are an infinite number of termination impedance combinations that can result in any given *standing wave ratio* (SWR). Devices such as RF amplifiers connected to the input end of a lossless transmission line will see the same SWR regardless of line length, but the *impedance* seen will be a function of the length of the line. This *impedance transformation* property may be used by designers to couple the input of an RF device to its source, or the output of a device to its load. At different points along a transmission line at increasing distances from a mismatched load, the impedance changes continuously until a point 1/2-wavelength from the load is reached. Here, the input impedance is the same as the load impedance. In between, the impedance changes from purely resistive (at two dis-

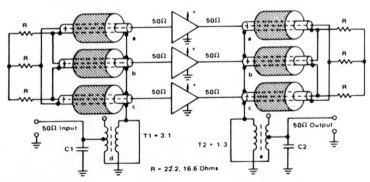

Figure 5-18 Amplifier circuit built around a three-port hybrid splitter/combiner. *(Courtesy of Motorola.)*

tinct values) to capacitive, then to inductive, and back again to the original impedance at the 1/2-wavelength point. This process repeats itself every 1/2-wavelength. A point 1/4-wavelength from the load appears as almost a mirror image of the load impedance. A capacitive load looks inductive; a short circuit looks like an open.

Through proper use of these properties, simple and efficient coupling circuits can be produced. These circuits can be formed using sections of miniature transmission line attached to the power amplifier circuit board, or simply as traces above a ground plane on the printed wiring board itself. The later case is referred to as *stripline coupling*. Figure 5-19 shows a single-stage RF amplifier utilizing stripline and discrete capacitors for input and output coupling. This technique is popular with equipment designers because of the following benefits:

- *Low cost.* Stripline coupling networks are simply a part of the PC board layout. No assembly time is required during construction of the system.
- *Excellent repeatability.* Variations in dimensions, and therefore performance, are eliminated.

Stripline has the following drawbacks:

- *Potential for radiation.* Depending on the design, shielding of stripline sections may be necessary to prevent excessive RF emissions. Coaxial cable-based coupling networks are not subject to this problem.
- *Repair difficulties.* If a stripline section is damaged from a lightning strike to the antenna, for example, it may be necessary to

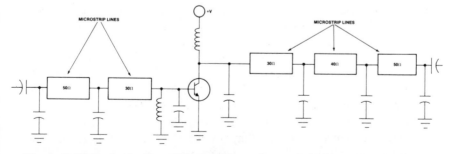

Figure 5-19 A single-stage RF amplifier using stripline coupling in the input and output circuits. (*Source: Mobil Radio Technology Magazine,* an Intertec Publication.)

replace the entire PWB. In a coaxial cable-based network, one or more pieces of cable may have to be replaced, but the circuit board should survive.

Stripline Construction Techniques. Stripline technology typically utilizes a double-sided printed circuit board made of fiberglass. The board is usually 30- to 50-thousandths of an inch thick. The board is uniform over the entire surface and forms an electrical ground plane for the circuit. This ground plane serves as a return for the electrical fields built up on the component side of the board.

The shape and length of each trace of stripline on the component side dictates the impedance and reactance of the trace. Whereas traces on a regular printed circuit board merely serve to connect components together, traces on a stripline board not only connect, but replace many passive devices, such as coils, transformers, resistors, and capacitors.

The impedance of a trace is a function of the width of the trace, its height above the lower surface ground plane, and the dielectric constant of the circuit board material. The length of the trace is another important factor. At microwave frequencies, a 1/4-wavelength can be as short as 0.5-in in air. Because all printed circuit boards have a dielectric constant that is greater than the dielectric constant of air, waves are slowed as they travel through the board/trace combination. This effect causes the wavelength on a circuit board to be dependent on the dielectric constant of the material of which the board is made. At a board dielectric constant of 5 to 10 (common with the materials typically used in printed circuit boards), a wavelength may be up to 1/3 shorter than when in air. The wider the trace, the lower the RF impedance.

Traces that supply bias and require operating or control voltages are usually made thin so as to present a high RF impedance while maintaining a low dc resistance. Narrow bias and control traces are usually made to be a multiple of 1/4-wavelength at the operating frequency so unwanted RF energy may be easily shunted to ground.

Figure 5-20 shows stripline serving several functions in a satellite-based communications system. The circuit includes: a 3-section, low-pass filter; a 1/4-wave line used as half of a transmit/receive switch; bias lines to supply +5 Vdc to a transistor; impedance-matching strip segments that convert a high impedance (130 Ω) to 50 Ω; and coupling lines that connect two circuit sections at one impedance.

Application Example. An example circuit using stripline design techniques is shown in Figure 5-21. The 25 W UHF amplifier (fre-

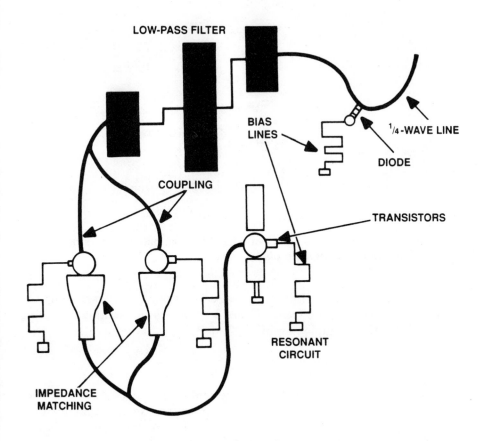

Figure 5-20 A typical application of stripline showing some of the components commonly used. (Source: Broadcast Engineering Magazine.)

quency 450-512 MHz) is designed for operation at 12.5 V. Stripline was chosen because of its inherent superiority over other methods at high frequencies. Stripline construction is more efficient than lumped constant equivalents and is less expensive to build. A Teflon-bonded fiberglass dielectric PC board is used in the design. (Teflon is a registered trademark of DuPont.) A substrate thickness of 1/16-in., coupled with a trace line width of 0.22-in. (the same width as the transistor leads), produces a characteristic impedance of approximately 40 Ω. A variety of techniques may be used to synthesize the coupling networks. There are, in fact, an infinite number of solutions to a given coupling design. After an initial choice of one of the components is made, however, only a small number of solutions are practical. While it is apparent that all components must be kept within

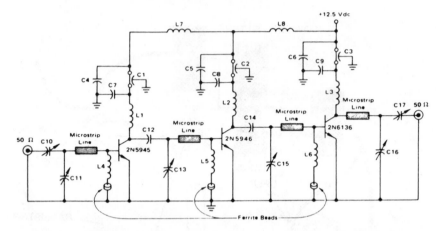

Figure 5-21 Schematic diagram of a 25 W UHF amplifier using stripline coupling. (Courtesy of Motorola.)

reasonable physical limits, the most critical parameter is usually the length of the stripline trace. Figure 5-22 shows the stripline coupling layout for the example amplifier.

5.2.3 Output Networks

The design of a network to resonate and couple the output of an RF transistor or FET to its load is a complicated exercise that involves balancing a number of parameters. Amplifiers that operate class C rely on a tank circuit to resonate the stage.

Consider an amplifier designed to deliver 50 W into a 50 Ω load, working with a 12 V power supply. A pair of L networks is used for impedance coupling. The basic equation $E = \sqrt{PR}$ shows that to deliver 50 W into a 50 Ω load, 50 V RMS must be supplied. In order to

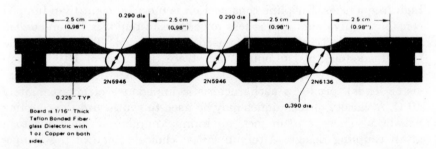

Figure 5-22 Physical layout of the 25 W UHF amplifier shown in the previous figure. (Courtesy of Motorola.)

get 50 V RMS from a 12 V supply, the output network must act as a step-up transformer. Stated from a different perspective, the output network must transform the load so the transistor sees a lower impedance.

Suitable circuits include transformers, Pi networks, T networks, and L networks. The approach chosen depends on the operating frequency, preference of the designer, and the power level. If the network has reasonable Q, the waveform of Figure 5-23 will be observed at the transistor collector when the amplifier is driven to full output. The output will swing down from the supply voltage to a minimum point (E_{sat}), limited by the transistor, and will swing above the supply voltage by approximately the same amount. This increased voltage is a result of the *flywheel effect* of a resonant circuit. In order to achieve the desired 50 W output, the coupling network must transform the 50 Ω load to present 1.0 Ω to the transistor.

This greatly simplified example brings up an important point in output coupling circuit design: seldom are output networks designed to *match* the output impedance of a transistor to the external load. The "equivalent circuits" concept states, in part, that *maximum power transfer* occurs when the load is *matched* to source (see Figure 5-24). The key point is that maximum power is not always wanted. Output networks are designed and adjusted to the impedance that will result in the desired power output and bandwidth. There can be a big difference between maximum power output from a stage, and the operating power that will provide the required parameters and not destroy the output device.

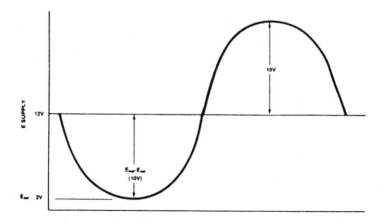

Figure 5-23 Collector waveform exhibited by a tuned RF amplifier. (*Source: Mobil Radio Technology Magazine,* an Intertec Publication.)

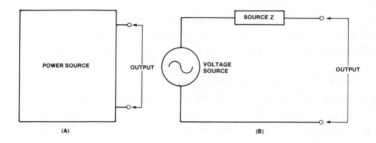

Figure 5-24 The "equivalent circuits" concept: a power source (a), no matter how complex, can be represented by a simple equivalent circuit (b) consisting of a voltage source with a series impedance. (*Source: Mobil Radio Technology Magazine,* an Intertec Publication.)

It should be obvious that just because the output impedance of the transistor is not matched, neither is it ignored. Choosing a device for a given application involves the following criteria:

• Available supply voltage
• Gain required at the operating frequency
• Output current required
• Thermal dissipation at the operating power level

The device may also have some reactance that can be used to its advantage in the coupling network.

Stray Reactances. As the operating frequency increases, stray inductance and capacitance must be accounted for in the design of input, interstage, and output coupling networks. These reactances present a number of problems to circuit designers. They may also, however, offer simple solutions to difficult coupling designs.

To illustrate the point, consider again the 1 Ω double L network discussed previously. Calculated for a maximum operating frequency of 150 MHz, the input inductor value would be about 3.0 nH. This is a pretty small inductance. It represents, in fact, just a fraction of an inch of printed wiring board trace. This raises the question: Where does the transistor lead end and the inductor begin? It is clear that at VHF frequencies and above, keeping the physical size of the device and PWB traces small are important considerations. A clever designer may, however, be able to incorporate the stray inductance of component leads or circuit board traces to an advantage, eliminating certain discrete components. This technique has been applied commonly in cavity-based vacuum tube RF amplifiers.

Fixed Frequency Circuits. One of the often-mentioned attributes of a solid state amplifier is its wide operating bandwidth. Indeed, this point has been discussed more than once so far in this book. It is only fair to point out, however, that broadband operation for solid state systems is not really cleverness at all. Using semiconductor devices means working at much lower voltages than vacuum tubes. Low voltage means low impedance, and low impedance translates to high-value capacitors and low-value inductors in coupling networks. With these design criteria, it often becomes impractical to make components variable. As a result, fixed networks are used. The result is amplifiers that do not have to be tuned. Or, stated a bit differently, amplifiers that cannot be tuned. Solid state amplifiers also typically are more sensitive to changes in load impedance. This fact must be taken into account in the design of transmission line, filter, and antenna systems.

Output Impedance. It is sometimes stated that the output impedance of an RF amplifier is 50 Ω. While this may occasionally be the case, more often than not the *output impedance*, as measured from the output terminals, is some other value. To state the case correctly, it should be said that the amplifier is designed to *work into* a 50 Ω load.

5.2.4   Practical Applications

High power transmitters based both on bipolar and FET devices are common today. Most new development, however, appears at this writing to be focused on MOSFET technology. In medium wave AM applications, FETs are typically used in a *switched bridge* or "H" configuration. Efficiencies of 95 percent or more have been reported. Such efficiencies result in minimal thermal stress on the devices, and thus long life. FETs are also used at VHF frequencies, but at lower power levels (75-100 W). High power devices, such as 500 W, are not widely used because of thermal management problems. The life of a semiconductor is a function of its operating junction temperature. Smaller chips tend to spread the heat better over the junction, reducing thermal stress.

The operating efficiency of FET devices at VHF frequencies for FM service has steadily improved, but still lags behind the best vacuum tube PAs. It is not uncommon for a tetrode to achieve 80 percent efficiency, including output filter losses. FET-based systems for FM typically operate with efficiencies between 70 and 79 percent. Efficiency is limited because of losses in the combining and filtering networks.

Few power FET devices are available for operation at UHF frequencies at costs that compare favorably with bipolar transistors. Work continues, however, to improve FET performance at high frequencies, where bipolar devices are now commonly found.

Switching RF Amplifier. A variety of amplifier designs based on switching systems have been developed to improve the efficiency performance of medium- and high-power transmitters. A power amplifier module for medium wave AM is shown in simplified form in Figure 5-25 (a). This method of generating a modulated RF signal employs a solid-state switch to swing back and forth between two voltage levels at the carrier frequency. The result is a square-wave signal that is filtered to eliminate all components except the fundamental frequency itself. A push-pull version of the circuit is shown in Figure 5-25(b). This approach is adequate if you are only interested in generating a simple sinusoid at the carrier frequency.

Figure 5-26 illustrates a class D switching system utilizing bipolar transistors that permits the generation of a modulated carrier. The

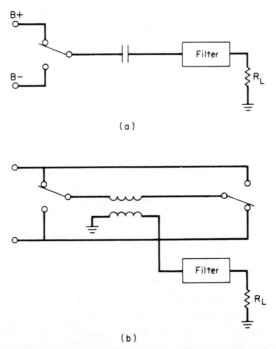

Figure 5-25 The basic concept behind class D radio frequency amplifiers for MF service: (a) single-ended circuit; (b) push-pull circuit. (*Source:* K. Blair Benson, *Audio Engineering Handbook,* McGraw-Hill, New York, 1988.)

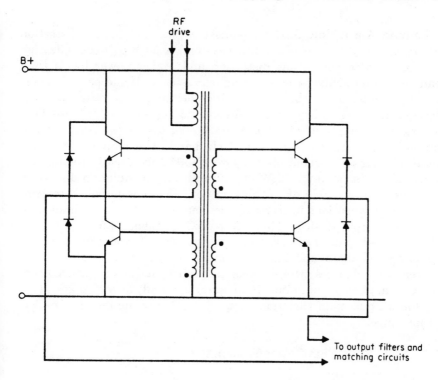

Figure 5-26 Bipolar transistor RF stage operating in a class D switching mode for MF service. (*Source:* K. Blair Benson, *Audio Engineering Handbook,* McGraw-Hill, New York, 1988.)

dc supply to the RF amplifier stages is switched on and off by an electronic switch in series with a filter. Operating in the class D mode results in a composite signal similar to that generated by the vacuum tube class D amplifier in the PDM system (at much higher power). (The PDM process is described in Chapter 2.) The individual solid state amplifiers are typically combined through a toroidal filter. The result is a group of low-powered amplifiers operating in parallel and combined to generate the required energy.

Transmitters up to 50 kW and above have been constructed using this design philosophy. They have proved to be reliable and efficient. The parallel design provides users with automatic redundancy and has ushered-in the era of "graceful degradation" failures. In almost all solid-state transmitters, the failure of a single power device or stage will reduce the overall power output or modulation capability of the system, but will not take the transmitter off the air. The negative effects on system performance are usually negligible.

Linear Amplifier. SSB transmission places stringent distortion requirements on a power amplifier system. A push-pull configuration can be used to obtain low distortion and higher power levels than can be readily achieved with a single transistor. While parallel operation can often meet the power output demands, the push-pull configuration offers improved even-harmonic suppression. Figure 5-27 shows an example of an 160 W *peak envelope power* (PEP) linear amplifier operating in the 3-30 MHz band. The amplifier offers a total power gain of about 30 dB. Two 2N5942 are used in the design. Each device is rated for 80 W output with intermodulation distortion (IMD) products at -30 dB. A biasing adjustment is provided to compensate for variations in transistor current gain. A 2N6370 is used to drive the output devices. The driver produces about 4.5 W (PEP) for the push-pull pair. Transmission line-type transformers are used for coupling and signal-splitting. To compensate for variations in output power with changes in operating frequency, negative voltage feedback is applied to both the final amplifier and driver stages. This feedback limits gain variation to just 0.5 dB over the 3-30 MHz operating range.

5.3 Solid State High-Power Transmitter

Until recently, AM broadcast transmitters generally relied on only a couple of basic designs. Digital modulation techniques, however, offer

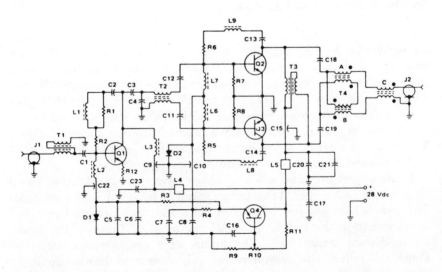

Figure 5-27 160 W (PEP) broadband linear amplifier. *(Courtesy of Motorola.)*

new ways to significantly increase the operating efficiency and audio performance of a transmitter. Digital modulation may be implemented through a variety of schemes. The approach described in this section is based on the DX-10 AM transmitter (built by Harris). A simplified diagram of the transmitter is shown in Figure 5-28.

The RF system involves four basic steps:

- The modulating audio signal is low-pass filtered and a dc control voltage is added. This audio processing stage provides an output signal that determines both the carrier level and modulation level.
- The composite audio + dc signal is digitized into a 12-bit word data stream. This analog-to-digital (A/D) conversion takes place in a high-speed converter. The audio is sampled at the carrier frequency or a sub-multiple of the carrier frequency.
- A modulation encoder codes the data into on-off signals for the RF power amplifier.
- Signals from the modulation encoder switch individual RF power amplifiers on and off as needed to produce an amplitude modulated signal.

An AM waveform has a constantly changing RF level. In a digital AM transmitter, the RF level at the output is varied by switching on either a larger or smaller number of RF power amplifiers. Based on signals from the modulation encoder, the correct number of RF amplifier modules are turned on or off. The outputs of these amplifiers are then combined to produce the composite AM signal.

The RF power amplifier consists of 48 modules in a 10 kW system. Because 48 equal steps is not enough to accurately reproduce the modulation envelope required, binary weighted steps are used to obtain the required resolution. Forty-two of the amplifiers are assigned

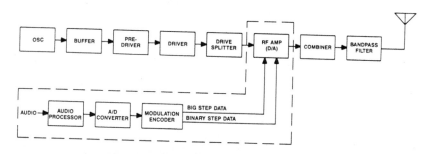

Figure 5-28 Block diagram of a digital AM transmitter. (*Source: Broadcast Engineering Magazine.*)

so-called *big steps*. Each big step amplifier contributes an equal voltage to the combined output.

The *binary steps* (bits 7, 8, 9, 10, 11 and 12) are controlled directly by the A/D converter and contribute 1/2, 1/4, 1/8, 1/16, 1/32, and 1/64 the voltage of the big steps, respectively.

The RF amplifiers are essentially constant voltage sources. The outputs are combined in a multi-turn primary, single-turn secondary, ferrite combiner to form an amplitude-modulated carrier with a quantized envelope. The quantized AM waveform is then filtered by a bandpass network to remove unwanted spectral components.

5.3.1   Audio Processing

The audio processor, shown in Figure 5-29, contains a low-pass Bessel filter, which attenuates frequencies above the required audio frequency range. The sampling rate is very high, determined by the operating frequency. The filter cutoff point need be only approximately 250 kHz, or one-half the operating frequency. This high cutoff frequency eliminates the need for sharp cutoff filters.

At any instant, transmitter power level is dependent upon the number of RF amplifiers that are turned on. Therefore, if the audio signal is removed, the transmitter power level would theoretically drop to zero. The unmodulated, maximum power level is set using a summing circuit consisting of a differential amplifier with the audio signal at one input and a negative dc voltage at the other. This small negative voltage determines the unmodulated carrier output level.

The audio feeds an analog multiplier, configured to operate as a divider. The divider develops a control voltage that maintains constant transmitter power output with varying supply voltages. The audio signal is applied to one input of the divider and a sample of

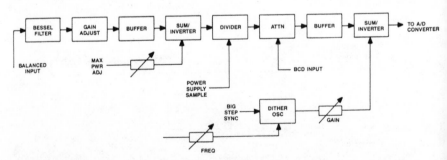

Figure 5-29 Audio processing section of the digital AM transmitter. The processor develops the audio + dither + dc control voltage signal, which is then digitized. (*Source: Broadcast Engineering Magazine.*)

the high-voltage power supply (+230 Vdc) is applied to the other input. If the +230 Vdc changes, which could cause the RF power output to change, the divider changes the audio + dc level to compensate for the variation.

A *dither* signal is added to the audio + dc signal to reduce noise that could result from the A/D conversion process. The dither signal is a high-frequency triangle wave, inserted at approximately 60-70 dB below the nominal audio level.

The dither oscillator receives a big step synchronizing signal from the A/D converter, which synchronizes the dither with the switching of the big steps. This synchronization "forces" a decision in the A/D converter when that stage is faced with low-level audio signals. The dither signal level represents approximately the size of the 12th bit of resolution. This process results in an effective audio resolution of 13 to 14 bits.

The dc + audio signal is also used to help regulate the B- supply voltage. The dc + audio modulates the B- regulator output, thereby effectively providing a *dynamic bias* voltage for the RF power transistors. The technique improves the switching performance of the transistors when they are operating under high current conditions.

5.3.2   A/D Converter

The A/D converter, shown in Figure 5-30, takes an analog signal (with a dc component to control carrier power level) and converts it to a digital signal consisting of a 12-bit binary word. The A/D converter section also contains circuitry to provide:

- Logic control signals (derived from an RF sample) for the A/D conversion process
- Synchronization of the 12-bit digital word with the RF carrier signal at the power amplifiers so that power stages switch on and off at the proper time
- A conversion error logic signal for the status indicators
- A big step sync signal to insure that unwanted steps or glitches (noise) in the modulated output signal do not occur
- 12-bit D/A conversion that provides an audio signal to the envelope error detector circuit in the controller

The A/D board analog input stage inverts the audio + dc signal and adds a small/big step sync component. This added component ensures that when a big step amplifier switches on or off, it does not change state again and cause a transient or glitch in the RF output.

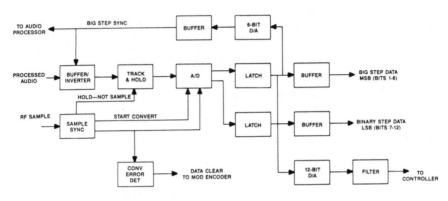

Figure 5-30 A/D converter subsection. The RF sample is used to synchronize the track-and-hold and A/D converter, which produce the big-step and binary-step data. (*Source: Broadcast Engineering Magazine.*)

The A/D conversion process requires two steps in less than 1 μs. The analog input signal (audio + dc) is sampled and stored by a track and hold circuit, and then fed to the input of the A/D converter. The function of the track and hold is to ensure that the signal input to the A/D converter remains constant while the conversion process is underway.

The *sample sync* circuit takes an RF sample from the power amplifier RF drive splitter, divides it by 1, 2, or 3, depending on the transmitter operating frequency, and generates control signals to start and complete the conversion process.

The A/D sampling process generates a digital audio spectrum at baseband and replicates this spectrum at multiples of the sampling frequency. Selection of the sample rate is important to ensure that all unwanted spectral components fall outside the filter passband. The sample rate, determined by a divider in the sample sync circuit, ranges between 400 kHz and 820 kHz, depending on the operating frequency.

The *end-of-conversion* logic signal from the A/D converter provides data strobe signals to latch-on the A/D and modulation encoder boards. Each time another conversion is performed, the end-of-conversion output is strobed, and the output data is latched until the next conversion. If the A/D conversion process does not take place because of some fault, a conversion error circuit provides a *data clear* logic signal to clear the latches. All the power amplifier stages are thus turned off.

5.3.3 Modulation Encoding

The modulation encoder, shown in Figure 5-31, accepts the 12-bit digital audio information and encodes it into on-off signals for the RF power amplifier stages. The 12-bit signal from the A/D converter is first stored in latches.

For the six least significant bits (bits 7 to 12), the data latch outputs provide PA on-off control signals through latches, OR gates and drivers to six binary amplifiers. The data latches for the most significant bits (bits 1 to 6) address read only memory (ROM) chips.

Based on the input signals supplied by the latches, a binary address, or location, within the ROM is identified and a predetermined sequence of logic high and low signals appears at the ROM output pins. The ROM outputs are first stored in data latches, then buffered before driving the 42 big step RF amplifiers. Jumpers are used to connect the latch outputs to the buffer inputs. This permits patching around a failed power amplifier stage. Should an RF stage fail, it is possible to substitute another amplifier. The loss of one amplifier increases distortion to approximately two percent. Substituting one of the amplifiers used only during modulation peaks reduces both the distortion and available peak power.

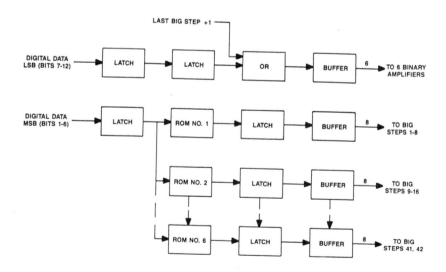

Figure 5-31 Modulation encoder subsection. The encoder develops the control signals required by the binary- and big-step amplifiers. (*Source: Broadcast Engineering Magazine.*)

Table 5-2 ROM decoder logic table. When addressed by the modulation encoder, the ROMs develop the necessary drive signals for the power amplifier. (*Source: Broadcast Engineering Magazine.*)

	BIT NO.								BIG STEP							
	1	2	3	4	5	6	7	8	1	2	3	4	5	6	7	8
ROM NO. 1	0	0	0	0	0	0	X	X	0	0	0	0	0	0	0	0
	0	0	0	0	0	0	X	X	0	0	0	0	0	0	0	0
	0	0	0	0	0	0	X	X	0	0	0	0	0	0	0	0
	0	0	0	0	0	0	X	X	0	0	0	0	0	0	0	0
	0	0	0	0	0	1	X	X	1	0	0	0	0	0	0	0
	0	0	0	0	1	0	X	X	1	1	0	0	0	0	0	0
	0	0	0	0	1	1	X	X	1	1	1	0	0	0	0	0
	0	0	0	1	0	0	X	X	1	1	1	1	0	0	0	0
	0	0	0	1	0	1	X	X	1	1	1	1	1	0	0	0
	0	0	0	1	1	0	X	X	1	1	1	1	1	1	0	0
	0	0	0	1	1	1	X	X	1	1	1	1	1	1	1	0
	0	0	1	0	0	0	X	X	1	1	1	1	1	1	1	1

001001XX TO 111111XX   YIELDS 11111111

	BIT NO.								BIG STEP							
	1	2	3	4	5	6	7	8	9	10	11	12	13	14	15	16

000000XX TO 001000XX   YIELDS 00000000

	1	2	3	4	5	6	7	8	9	10	11	12	13	14	15	16
ROM NO. 2	0	0	1	0	0	1	X	X	1	0	0	0	0	0	0	0
	0	0	1	0	1	0	X	X	1	1	0	0	0	0	0	0
	0	0	1	0	1	1	X	X	1	1	1	0	0	0	0	0
	0	0	1	1	0	0	X	X	1	1	1	1	0	0	0	0
	0	0	1	1	0	1	X	X	1	1	1	1	1	0	0	0
	0	0	1	1	1	0	X	X	1	1	1	1	1	1	0	0
	0	0	1	1	1	1	X	X	1	1	1	1	1	1	1	0
	0	1	0	0	0	0	X	X	1	1	1	1	1	1	1	1

010001XX TO 111111XX   YIELDS 11111111

ROM Coding. The modulation encoder contains six identical ROMs. Each ROM is 8-by-8 (8 binary inputs and 8 logic outputs) with 28 possible addresses. (Refer to Table 5-2 for the following description.)

While bits 1 through 8 from the A/D converter are required to step through all the ROM addresses, only bits 1 through 6 are used to select the turn-on signals for the 42 big steps. The X's shown in Bit columns 7 and 8 indicate a "don't care" value and are used to simplify the explanation of the ROM table. Before each new big step is turned on by a status output change from 0 to 1, bits 7 and 8 clock through the sequence of 00, 01, 10, and 11. This sequence is shown only once with X's on the first four lines at top left of table for simplicity. As bits 3 through 6 for ROM #1 are clocked through binary 1 through 8, the bit-step control outputs change from 0 to 1 for big steps 1 through 8. The binary input to ROM #1, after reaching 001000XX, (big steps 1 through 8 and higher) continues to 111111XX. (All ROM's receive the same binary data.) The first eight big steps are therefore kept on for all binary inputs above the count of 8.

ROM #2, which controls big steps 9 through 16, does not turn on any big steps until the binary count of 9 is reached. From binary 9 through binary 16, ROM #2 turns on big steps 9 through 16 then keeps them on through binary 256 (111111XX). The remaining four ROM's continue turning on big steps in the same manner until all 42 big steps are active. If the modulating signal continues to increase after all available big steps are turned on, the binary steps would continue to turn on and off, causing a sawtooth to appear on the top of the clipped waveform. This is prevented by feeding the last big step +1 logic turn-on signal to the input of the 6 OR gates. This signal turns all of the binary steps and holds them there until the peak has passed.

The six binary steps can contribute $(2^6) - 1$ steps or 63 different power levels. Before each big step is turned on, the binary steps must go through the entire count from one to 63. With all 42 big steps active, the total number of levels traversed becomes the number of big steps times the number of discrete levels produced by the six binary amplifiers, or, $42 \times 63 = 2646$.

After the last big step has turned on, the binary steps count through one more sequence to 63. There are also 42 individual big steps, so the total available steps becomes $2646 + 63 + 42$, or 2751. This is approximately equivalent to 11.43 bits of resolution ($211 \times 43 = 2750$). The waveform developed by switching on the big steps and binary amplifiers is shown in Figure 5-32.

5.3.4 D/A Conversion

The digital information, generated by the A/D converter, switches units of RF voltage on or off by switching RF amplifiers on and off. The RF power amplifier may be thought of as a D/A converter, where the digital input signal is converted to a high power, amplitude modulated (analog) RF signal.

The power amplifiers produce equal RF voltage steps (not equal RF power steps) at the combiner output (see Figure 5-33). The PA amplifier output power depends on the total number of stages that

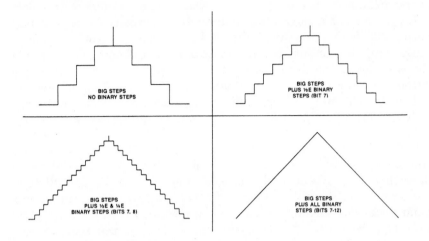

Figure 5-32 The combining process for the big- and binary-step output signals. (*Source: Broadcast Engineering Magazine.*)

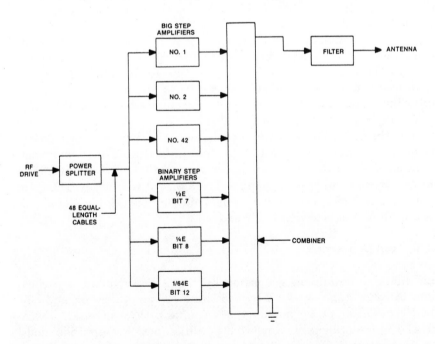

Figure 5-33 Power amplifier subsection. The PA comprises 42 big-step and 6 binary-step amplifiers. (*Source: Broadcast Engineering Magazine.*)

are switched on. Switching on twice as many RF amplifier stages will produce twice the voltage output and, therefore, four times the output power.

The binary-weighted voltage contribution of the six binary amplifiers is obtained by the proper selection of supply voltage, number of output transformer primary turns, or a combination of both. A transmitter with a carrier power of 10 kW produces a peak output power of 57.6 kW at 140% positive modulation. In a full 12-bit binary sequence D/A converter, the RF amplifier corresponding to the most significant digit in the digital word must deliver an output of 43.2 kW. The next most significant digit would correspond to 10.8 kW. Because such large amplifiers are not only more difficult to build, but difficult to switch on and off quickly, the transmitter design uses binary-weighted amplifiers only at the low output levels (up to 1.34 kW). Figure 5-34 illustrates the power levels that would be required in both the 12-bit binary and pseudo-binary (42 unity + 6-bit binary) configurations. Note that in the later configuration, the maximum power required from a single RF amplifier is 1.34 kW, as compared to the 43.2 kW required in the 12-bit binary configuration.

Power Amplifier. The transmitter uses class D switching amplifiers, each with four power MOSFETS operating in a bridge configuration, as shown in Figure 5-35. The RF drive to each power MOSFET is divided by separate secondary windings on T1 and T2.

Back-to-back zener diodes CR1, CR2, CR3, and CR4 provide protection for the MOSFET inputs from transients and excessive drive. Each power MOSFET in the bridge acts like a switch and is either on or off at any particular time. Devices Q1 and Q4 are driven 180° out of phase with Q2 and Q3.

The amplifier output is coupled to the output combiner by T3. During one half of the RF cycle, Q1 and Q4 are driven into saturation, while Q2 and Q3 are in cutoff. During the other half of the RF cycle, Q2 and Q3 are driven into saturation, while Q1 and Q4 are in cutoff. This produces a square wave peak-to-peak voltage of twice the bridge voltage (DC supply voltage) across the output transformer winding.

When the control signal is in the logic 0 state, the amplifier is turned off by the modulation circuit and does not deliver any power. The transformer's primary circuit is, however, essentially complete. The drive signal for the top two devices is still present. Current is allowed to flow through the active MOSFETs, the transformer winding, and the reverse diode of the other top devices. This design prevents the combiner from becoming an open circuit when the amplifier is turned off.

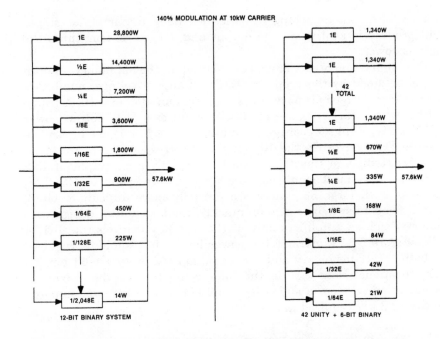

Figure 5-34 A true binary implementation of digital modulation would require an amplifier capable of 28.8 kW. The pseudo-binary design requires a maximum of 1.34 kW from any single amplifier. (*Source: Broadcast Engineering Magazine.*)

The modulation control circuit turns the RF amplifiers on and off, as directed by the encoder. Amplifiers are turned off by removing the RF drive to the bottom two devices (Q2 and Q4). When the input to the modulation section is approximately one volt or more positive, the control circuit shorts the RF drive signals to Q2 and Q4 to ground through diodes CR5 and CR6. The RF MOSFETs (Q2 and Q4) do not conduct, and therefore no current can flow through the combiner transformer primary winding. If the input voltage to the modulation stage is negative, the short is removed allowing RF drive to flow to the gates of PA devices Q2 and Q4.

5.3.5   Performance Benefits

The digital modulation scheme described in this section offers a significant improvement in operating efficiency. The ac-to-RF efficiency is approximately 86 percent, compared to the 60 to 70 percent efficiency of a comparable conventional transmitter. The approximate

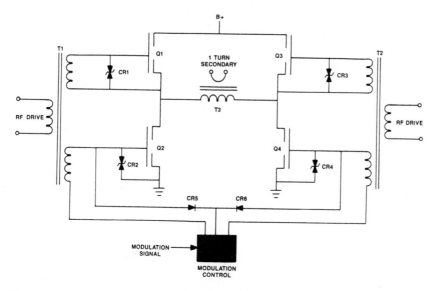

Figure 5-35 Schematic diagram of the power amplifier module. (*Source: Broadcast Engineering Magazine.*)

power consumption for the DX-10 is 17.5 kW. A similar 10 kW tube-based transmitter consumes approximately 28 kW. Table 5-3 lists typical performance specifications for the DX-10.

As of this writing, the digital modulation technique described here has been applied successfully to a 100 kW AM transmitter. The manufacturer plans to use the "building block" approach characterized by this technique for higher-power transmitters as well.

Bibliography

1. "High Power Transistor Replaces Vacuum Tubes in any Size of FM Transmitter," Philips Press Publication, Eindhoven, The Netherlands, December 1987.
2. Glen, Fred, "Don't Overlook Efficiency in RF Power Amplifier Systems," *Mobile Radio Technology*, Intertec Publishing, Overland Park, KS, December 1986.
3. Gutsche, Manny, "RF Power Amplifier Installation and Troubleshooting," *Mobile Radio Technology*, Intertec Publishing, Overland Park, KS, December 1986.
4. Peters, Daniel, "Some Common Myths About RF Power Amps," *Mobile Radio Technology*, Intertec Publishing, Overland Park, KS, December 1986.

Table 5-3 Typical performance of a 10 kW digital AM transmitter. (*Source: Broadcast Engineering Magazine.*)

Audio frequency response	+/-0.0 dB to -0.6 dB 20 Hz to 10 kHz
Harmonic distortion	0.43% @ 95% modulator 20 Hz to 10 kHz
I.M. Distortion	0.42% 4:1 SMPTE
Square Wave Overshoot	0.8% @ 100 Hz @ 95%
Modulation Tilt	0.8% @ 100 Hz @ 90%
Modulation carrier shift	-0.6%
AM hum/noise	-65 dB
Positive peak capability	+140%
IQM (average)	-42 dB @ 1 kHz @ 95%
IPM (average)	0.021 radians (1.2°)
PA efficiency	90%
Overall efficiency	86%

5. Fink, D., and D. Christiansen, *Electronics Engineer's Handbook*, Third Edition, McGraw-Hill Book, New York, 1989.
6. Jordan, Edward C., *Reference Data for Engineers: Radio, Electronics, Computer and Communications*, Seventh Edition, Howard W. Sams Company, Indianapolis, IN, 1985.
7. Dick, Brad, "Digital Modulation: DX-10 AM Transmitter," *Broadcast Engineering*, Intertec Publishing, Overland Park, KS, March 1988.
8. Swanson, Hilmer, "Digital AM Transmitters," *IEEE Transactions on Broadcasting*, Volume 35, Number 2, Washington, D.C., June 1989.
9. Richards, Nicholas, "DX-10 AM Transmitter," Harris Corporation technical report, Quincy, IL, 1989.
10. Becciolini, B., "Impedance Matching Networks Applied to RF Power Transistors," *Motorola RF Device Data Manual*, Application Note AN721, Motorola, Phoenix, 1988.
11. Young, Glenn, "Microstrip Design Techniques for UHF Amplifiers," *Motorola RF Device Data Manual*, Application Note AN548A, Motorola Semiconductor, Phoenix, 1988.
12. Granberg, Helge, "Broadband Linear Power Amplifiers Using Push-Pull Transistors," *Motorola RF Device Data Manual*, Application Note AN593, Motorola Semiconductor, Phoenix, 1988.

6

Power Grid Vacuum Tubes

6.1 Introduction[1]

The phrase, "high technology," is perhaps one of the more overused descriptions in our technical vocabulary. It is a phrase generally reserved for the discussion of integrated circuits, fiber optics, satellite systems, and the like. Few people would associate high technology with vacuum tubes. The notion that vacuum-tube construction is more *art* than *science* may have been true 10 or 20 years ago, but today it's a different story.

The demand on the part of industry for tubes capable of higher operating power and frequency, and the economic necessity for tubes that provide greater efficiency and reliability, have moved power tube manufacturers into the high-tech arena. Advancements in tube design and construction have given engineers new RF generating systems that allow industry to grow and prosper.

Power grid vacuum tubes have been the mainstay of transmitters since the beginning of radio. The need for advanced power tubes is being met today with new processes and materials. Users are asking for systems that incorporate solid-state components in low- and medium-power stages and vacuum tubes in high-power stages. Each technology has its place, and each has its strengths and weaknesses.

---

1 Portions of this chapter were adapted from *The Care and Feeding of Power Grid Tubes*, Varian Associates, San Carlos, CA, 1982.

6.1.1 Types of Vacuum Tubes

If you bring up the subject of vacuum tubes to someone who has never worked with RF equipment, you will usually get a blank stare and a question something like, "Do they make those anymore?" Receiving tubes have more or less disappeared from the scene because of the development of transistors and integrated circuits. But power grid tubes continue to push the limits of technology. What solid-state device can deliver 60 kW output at VHF frequencies with a drive of 800 W? Such a device does not exist. Power grid tubes are an important part of RF technology today.

A power-grid tube is a device using the flow of free electrons in a vacuum to produce useful work. It has an emitting surface (the cathode), one or more grids that control the flow of electrons, and an element that collects the electrons (the anode). Power tubes can be separated into groups according to the number of electrodes (grids) they contain. The physical shape and location of the grids relative to the plate and cathode are the main factors that determine the *amplification factor* and other parameters of the device. The physical size and types of material used to construct the individual elements determines the power capability of the tube. A wide variety of tube designs are available to industrial users. By far the most common are triodes and tetrodes.

Triode. The power triode is a three-element device commonly used in a wide variety of RF generators. Triodes have three internal elements: the cathode, control grid, and plate. Most tubes are cylindrically symmetrical. The filament or cathode structure, the grid, and the anode are all cylindrical in shape and are mounted with the axis of each cylinder along the center line of the tube.

Some triodes are manufactured with the cathode, grid and anode in the shape of a flat surface, as shown in Figure 6-1. Tubes so constructed are called *planar* triodes. This construction technique permits operation at high frequencies. The close spacing reduces electron *transmit time*, allowing the tube to be used at high frequencies (up to 3 GHz). The physical construction of planar triodes results in short lead lengths, which reduces lead inductance. Planar triodes are used in both CW and pulsed modes. The contacting surfaces of the planar triode are arranged for easy integration into coaxial and waveguide resonators. Placement of the control grid relative to the cathode and plate determines the amplification factor (μ) of the triode. The μ values of triodes generally range from 5-200. Key mathematical relationships include:

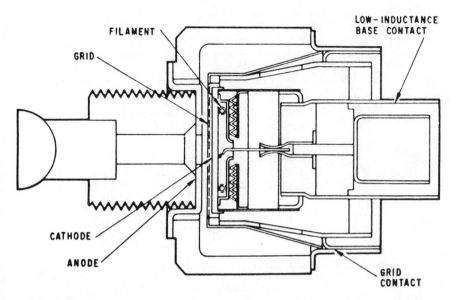

Figure 6-1 Internal configuration of a planar triode. *(Courtesy of Varian Associates.)*

- Amplification factor (μ) = $\Delta E_b / \Delta E_{c1}$ (with plate current held constant)
- Dynamic plate resistance (Rp) = $\Delta E_b / \Delta I_b$
- Transconductance (Sm or Gm) = $\Delta I_b / \Delta E_{c1}$

Where:
E_b = total instantaneous plate voltage
E_{c1} = total instantaneous control grid voltage
I_b = total instantaneous plate current

Figure 6-2 plots plate current and grid current as a function of plate voltage at various grid voltages for a triode with a μ of 20.

Triodes with a μ of 20-50 are generally used in conventional RF amplifiers and oscillators. High μ triodes (200 or so) may be designed so that the operating bias is zero (see Figure 6-3). These so-called *zero-bias triodes* are available with plate dissipation ratings of 400 W to 10 kW. The zero-bias triode is commonly used in grounded-grid amplifiers. The tube offers good power gain and circuit simplicity. No bias power source is required. Further, no protection circuits for loss of bias or drive are needed.

Low and medium μ devices are usually preferred for induction heating applications because of the wide variations in load that an

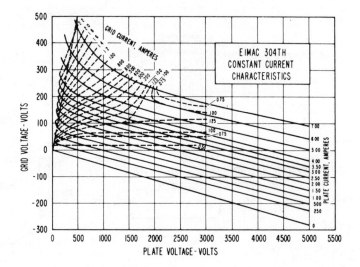

Figure 6-2 Constant current characteristics for a triode with a μ of 20. *(Courtesy of Varian Associates.)*

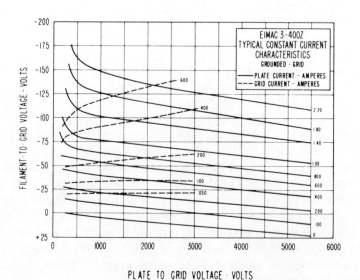

Figure 6-3 Grounded-grid constant current characteristics for a zero-bias triode with a μ of 200. *(Courtesy of Varian Associates.)*

induction or dielectric heating oscillator normally works into. Such tubes exhibit lower grid current variation with a changing load. The grid current of a triode with a μ of 20 will rise substantially less under a light or no-load condition than a triode with a μ of 40.

Vacuum tubes specifically designed for induction heating are available, intended for operation under adverse loading conditions. The grid structure is ruggedized with ample dissipation capability to deal with wide variations in load. As the load decreases, grid dissipation increases.

Tetrode. The tetrode is a four-element tube with two grids. The control grid serves the same purpose as the grid in a triode, while a second (screen) grid with the same number of vertical elements (bars) as the control grid, is mounted between the control grid and the anode. The grid bars of the screen grid are mounted directly behind the control grid bars as observed from the cathode surface, and serve as a shield or screen between the input circuit and the output circuit of the tetrode. The principal advantages of a tetrode over a triode include:

• Lower internal plate-to-grid feedback.
• Lower drive power requirements. In most cases, the driving circuit need supply only one percent of the output power.
• More efficient operation. Tetrodes allow for the design of compact, simple, flexible equipment with little spurious radiation and low intermodulation distortion.

Plate current is almost independent of plate voltage in a tetrode. Figure 6-4 plots plate current as a function of plate voltage at a fixed screen voltage and various grid voltages. In an ideal tetrode, there is no change in plate current with a change in plate voltage. The tetrode can, therefore, be considered a constant current device. The voltages on the screen and control grids determine the amount of plate current.

The total cathode current of an ideal tetrode is determined by the equation:

$$I_k = K \left[E_{c1} + \frac{E_{c2}}{\mu_s} + \frac{E_b}{\mu_p} \right]^{3/2}$$

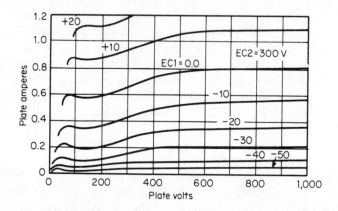

Figure 6-4 Tetrode plate current characteristics. Plate current is potted as a function of plate voltage, with grid voltages as shown. (*Source:* Fink and Christiansen, *Electronics Engineers' Handbook, 3rd Ed.,* McGraw-Hill, New York, 1989.)

Where:

I_k = cathode current
K = a constant determined by tube dimensions
E_{c1} = control grid voltage
E_{c2} = screen grid voltage
μ_s = screen amplification factor
μ_p = plate amplification factor
E_b = plate voltage

Pentode. The pentode is a five-electrode tube with three grids. The control and screen grids perform the same function as in a tetrode. The third grid, the *suppressor grid,* is mounted in the region between the screen grid and the anode. The suppressor grid produces a *potential minimum,* which prevents secondary electrons from being interchanged between the screen and plate. The pentode's main advantages include:

• Reduced secondary emission effects.
• Good linearity.
• Ability to let plate voltage swing below the screen voltage without excessive screen dissipation. This allows slightly higher power output for a given plate voltage.

Because of the design of the pentode, plate voltage has even less of an effect on plate current than in the tetrode. The same total space current equation applies to the pentode as with the tetrode:

$$I_k = K \left[E_{c1} + \frac{E_{c2}}{\mu_s} + \frac{E_b}{\mu_p} \right]^{3/2}$$

The suppressor grid may be operated at negative or positive with respect to the cathode. It may also be operated at cathode potential. It is possible to control plate current by varying the potential on the suppressor grid. Because of this ability, a modulating voltage can be applied to the suppressor to achieve amplitude modulation. The required modulating power will be low because of the low electron interception of the suppressor.

High Frequency Limits. Like most active devices, performance of a given vacuum tube deteriorates as the operating frequency is increased beyond its designed limit. Electron transit time is a significant factor in the upper frequency limitation of electron tubes. A finite time is taken by electrons to traverse the space from the cathode, through the grid, and on to the plate. As the operating frequency increases, a point is reached at which the electron transit-time effects become significant. This point depends on the accelerating voltages at the grid and anode, and their respective spacings. Tubes with reduced spacing in the grid-to-cathode region exhibit reduced transit-time effects.

There is also a power limitation that is interrelated with the high frequency limit of a device. As the operating frequency is increased, closer spacing and smaller-sized electrodes must be used. This reduces the power-handling capability of the tube. Figure 6-5 illustrates the relationship.

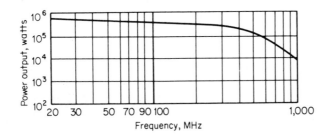

Figure 6-5 Continuous-wave output power capability of a gridded vacuum tube. (*Source:* Fink and Christiansen, *Electronics Engineers' Handbook,* 3rd Ed., McGraw-Hill, New York, 1989.)

Gridded tubes at all power levels for frequencies up to about 1 GHz are invariably cylindrical in form. At higher frequencies, planar construction is almost universal.

6.1.2   Thermal Emission from Metals

Thermal emission of electrons, *thermonic emission,* is the phenomenon of an electric current leaving the surface of a material as the result of thermal activation. Electrons with sufficient thermal energy to overcome the surface-potential barrier escape from the surface of the material. This thermally-emitted electron current increases with temperature because more electrons have sufficient energy to leave the material. Thermal electron emission can be increased by applying an electric field to the cathode. This field lowers the surface-potential barrier, and enables more electrons to escape. This field-assisted emission is known as the *Schottky* effect.

Secondary Emission. A material bombarded with energetic electrons will emit low-energy electrons called *secondary electrons.* The relationship of this property to the grid structures of a tube must be considered in any design. As the power capability of a vacuum tube increases, the physical size of the elements also increases. This raises the potential for secondary emission from the control, screen, and suppressor grids. Secondary emission can occur regardless of the type of cathode used. The yield of secondary electrons may be reduced through the application of surface treatments.

In a tetrode, the screen is operated at a relatively low potential, necessary to accelerate the electrons emitted from the cathode. Not all electrons pass through the screen on their way to the plate. Some are intercepted by the screen grid. As the electrons strike the screen, other low energy electrons are emitted. If these electrons have a stronger attraction to the screen, they will fall back to that element. If, however, they pass into the region between the screen and the plate, the much higher anode potential will attract them. The result is electron flow from screen to plate. Because of the physical construction of a tetrode, the control grid will have virtually no control over screen-to-plate current flow as a result of secondary electrons. During a portion of the operating cycle of the device, it is possible that more electrons will leave the screen grid than will arrive. The result will be a reverse electron flow on the screen element; this condition is often experienced in the operation of high power tetrodes. Because of this condition, a low impedance path for reverse electron flow must be provided.

6.2 Vacuum Tube Design

Vacuum tube technology is advancing just as fast as developments in solid-state technology, although accompanied by less fanfare. Power tubes today are designed with an eye toward high operating efficiency and high gain/bandwidth properties. Above all, a tube must be reliable and provide long operating life. The design of a new power tube is a lengthy process that involves computer-aided calculations and modeling. The design engineers must examine a laundry list of items, including:

- *Cooling*. How the tube will dissipate the heat generated during normal operation. A high-performance tube is of little value if it will not provide long life in typical applications. Design questions include whether the tube will be air-cooled or water-cooled, the number of fins the device will have, and the thickness and spacing of the fins.
- *Electro-optics*. How the internal elements line up to achieve the desired performance. A careful analysis must be made of what happens to the electrons in their paths from the cathode to the anode, including the expected power gain of the tube.
- *Operational parameters*. What the typical interelectrode capacitances will be, and the manufacturing tolerances that can be expected. This analysis includes: spacing variations between elements within the tube, the types of materials used in construction, the long-term stability of the internal elements and the effects of thermal cycling.

6.2.1 Device Cooling

The first factor that separates tube types is the method of cooling used: air, water, or vapor. *Air-cooled* tubes are common at power levels below 50 kW. A *water-cooling* system, while more complicated, is more effective than air-cooling (by as much as a factor of 5–10) in transferring heat from the device. Air-cooling at the 100 kW level is virtually impossible because it is difficult to physically move enough air through the device (if the tube is to be of reasonable size) to keep the anode sufficiently cool. *Vapor-cooling* provides an even more efficient method of cooling a PA tube than water-cooling. Naturally, the complexity of the external blowers, fans, ducts, plumbing, heat exchangers, and other hardware must be taken into consideration in the selection of a cooling method. Figure 6-6 shows how the cooling method is related to anode dissipation.

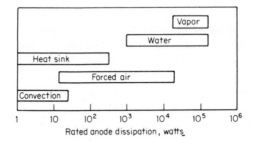

Figure 6-6 The relationship between anode dissipation and cooling method. (*Source:* Fink and Christiansen, *Electronics Engineers' Handbook, 3rd Ed.,* McGraw-Hill, New York, 1989.)

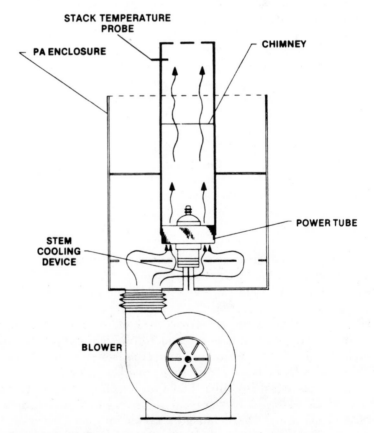

Figure 6-7 Typical transmitter PA stage cooling system. (*Source: Broadcast Engineering Magazine.*)

Air Cooling. A typical air-cooling system for a transmitter is shown in Figure 6-7. Cooling system performance for an air-cooled device is not necessarily related to airflow volume. The cooling capability of air is a function of its mass, not its volume. An appropriate airflow rate within the equipment is established by the manufacturer, resulting in a given resistance to air movement.

The altitude of operation is also a consideration in a cooling system design. As altitude increases, the density (and cooling capability) of air decreases. To maintain the same cooling effectiveness, increased airflow must be provided.

Water Cooling. Water cooling is usually preferred for power outputs above approximately 50 kW. Multiple grooves on the outside of the anode, in conjunction with a cylindrical jacket, force the cooling water to flow over the surface of the anode, as shown in Figure 6-8. Because the water is in contact with the outer surface of the anode, a high degree of purity must be maintained. A resistivity of 1 mΩ/cm (at 25°C) is typically specified by tube manufacturers. Circulating water can remove about 1 kW/cm^2 of the effective internal anode area. In practice, the temperature of the water leaving the tube must be limited to 70°C to prevent the possibility of spot boiling. After leaving the anode, the heated water is passed through a heat ex-

Figure 6-8 Water cooled anode with grooves for controlled water flow. *(Courtesy of Varian Associates.)*

changer where it is cooled to 30-40°C before being pumped back to the tube.

Vapor-Phase Cooling. Vapor cooling allows the permissible output temperature of the water to rise to the boiling point, enabling higher cooling efficiency compared with water cooling. The benefits of vapor-phase cooling are the result of the physics of boiling water. Increasing the temperature of one gram of water from 40°C to 70°C requires 30 calories of energy. However, transforming one gram of water at 100°C into steam vapor requires 540 calories. Thus, a vapor-phase cooling system permits essentially the same cooling capacity as water cooling, but with greatly reduced water flow. Viewed from another perspective, for the same water flow, the dissipation of the tube may be increased significantly (all other considerations being the same).

A typical vapor-phase cooling system is shown in Figure 6-9. A tube with a specially designed anode is immersed in a boiler filled with distilled water. When power is applied to the tube, anode dissipation heats the water to the boiling point, converting the water to steam vapor. The vapor passes to a condenser, where it gives up its energy and reverts to a liquid state. The condensate is then returned to the boiler, completing the cycle. Electric valves and interlocks are included in the system to provide for operating safety and mainte-

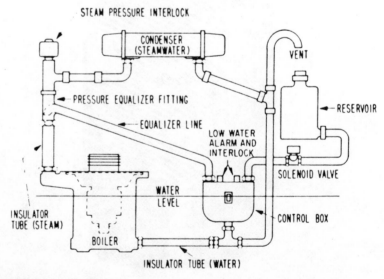

Figure 6-9 Typical vapor-phase cooling system. *(Courtesy of Varian Associates.)*

nance. A vapor-phase cooling system for a transmitter with multiple PA tubes is shown in Figure 6-10. In order to achieve the most efficient heat transfer, the anode must be structured to provide for optimum contact with the water in the boiler. Figure 6-11 shows several examples. The most common approach is to incorporate thick, vertical fins on the exterior of the anode to achieve a radial temperature gradient on the surfaces submerged in water. This way, hot spots do not cause thermal runaway of the tube. The temperatures of the fins vary from about 110°C at the tip to 180°C at the root. At low dissipation levels, boiling takes place at the root of the fin. As dissipation increases, the boiling point migrates outward.

Because the boiler is usually at a high voltage relative to ground, it must be insulated from the rest of the system. The boiler is usually mounted on insulators and the steam and water connections are made through insulated tubing. High voltage standoff is more easily accomplished in a vapor-phase system than in a water cooled system because:

• There is a minimum of contamination because the water is constantly being re-distilled.
• There is an inherently higher resistance in the system because of the lower water flow rate. The water inlet line is a smaller diameter, resulting in greater resistance.

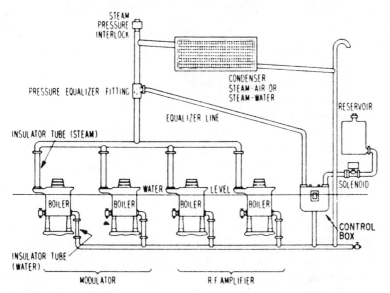

Figure 6-10 Vapor-phase cooling system for a 4-tube transmitter using a common water supply. *(Courtesy of Varian Associates.)*

Figure 6-11 Vapor-phase-cooled tubes removed from their companion boilers. *(Courtesy of Varian Associates.)*

In a practical system, a 2-ft. section of insulated tubing on the inlet and outlet ports is capable of a 20 kV standoff. A vapor-phase system is nearly 20 times more efficient than conventional water cooling. Because of the effects of *thermosyphoning*, natural circulation of the water eliminates the need for a pump.

The dramatic increase in heat absorption that results from converting hot water to steam is repeated in reverse in the condenser. As a result, a condenser of much smaller thermal capacity is required for a vapor-phase system as opposed to a water-cooled system. For example:

- In a practical water-cooled system, water enters the heat exchanger at 70°C and exits at 40°C; the mean temperature is 55°. For an ambient external temperature of 25°C, the mean differential between the water and the heat sink (air) is 30°C. The greater the differential, the more heat transferred.
- In a practical vapor-phase cooled system, water enters the heat exchanger as steam at 100°C and exits as water at 100°C; the mean temperature is 100°. For an ambient external temperature of 25°C, the mean differential between the vapor/water and the heat sink is 75°C.

In this example, the vapor-phase condenser is nearly three times more efficient than its water-cooled counterpart.

It should be noted that the condenser in a vapor-phase system may use either air or water as a heat sink. The water-cooled condenser provides for isolation of the PA tube cooling system from the outside world. This approach may be necessary because of high voltage or safety reasons. In such a system, the water is considered a *secondary coolant.*

Special Applications. Power devices used for research applications must be designed for transient overloading, requiring special considerations with regard to cooling. Oil, heat pipes, refrigerants (such as Freon) and, where high voltage holdoff is a problem, gases (such as sulfahexafluoride) are sometimes used to cool the anode of a power tube.

6.2.2   Cathode Assembly

The ultimate performance of any vacuum tube is determined by the accuracy of design and construction of the internal elements. The requirements for a successful tube include the ability to operate at high temperatures and withstand physical shock. Each element is critical to this objective.

The cathode used in a power tube obtains the energy required for electron emission from heat. The cathode may be directly heated (filament type) or indirectly heated. The three types of emitting surfaces most commonly used are:

- *thoriated tungsten*
- *alkaline-earth oxides*
- *tungsten barium aluminate-impregnated emitters*

The thoriated tungsten and tungsten-impregnated cathodes are preferred in power tube applications because they are more tolerant to *ion bombardment* (see the section "Ion Bombardment" that follows). The characteristics of the three emitting surfaces are summarized in Table 6-1.

A variety of materials may be used as a source of electrons in a vacuum tube. Certain combinations of materials are preferred, however, for reasons of performance and economics.

Oxide Cathode. The conventional production-type oxide cathode consists of a coating of barium and strontium oxides on a base metal such as nickel. The oxide layer is formed by first coating a nickel structure (a can or disc) with a mixture of barium and strontium carbonates, suspended in a binder material. The mixture is approximately 60 percent barium carbonate and 40 percent strontium carbonate. During vacuum processing of the tube, these elements are *baked* at high temperatures. As the binder is burned away, the car-

Table 6-1 Characteristics of common thermonic emitters.
(*Source:* Fink and Christiansen, *Electronics Engineers' Handbook, 3rd Ed.,*
McGraw-Hill, New York, 1989.)

Emitter	Heating method	Operating temp, °C	Emission density, A/cm^2	
			Average	Peak
Oxide	Direct and indirect	700–820	0.100–0.5	0.100–20
Thoriated tungsten	Direct	1600–1800	0.04–0.43	0.04–10
Impregnated tungsten	Direct and indirect	900–1175	0.5–8.0	0.5–12

bonates are subsequently reduced to oxides. The cathode is then *activated* and will emit electrons.

An oxide cathode operates CW at 700-820°C and is capable of an average emission density of 100-500 mA/cm^2. High emission current capability is one of the main advantages of the oxide cathode. Other advantages include high peak emission for short pulses and low operating temperature. As shown in the table, peak emission of up to 20 A/cm^2 is possible from an oxide cathode. A typical oxide cathode is shown in Figure 6-12.

Thoriated Tungsten Cathode. A thoriated tungsten filament is another form of an atomic-film emitter commonly used in power grid tubes. The thoriated-tungsten filament (or cathode) is created in a high-temperature gaseous atmosphere to produce a layer of *ditungsten carbide* on the surface of the cathode element(s). Thorium is added to tungsten in the process of making tungsten wire. Typically, concentration is about 1.5 percent thorium (in the form of thoria). By proper processing during vacuum pumping of the tube envelope, the metallic thorium is brought to the surface of the filament wire. The result is an increase in emission of approximately 1,000 times over a conventional cathode. At a typical operating temperature of 1,600-1,800°C, a thoriated tungsten filament will produce an average emission of 40-430 mA/cm^2. Peak emission ranges up to 10 A/cm^2.

One of the advantages of a thoriated tungsten cathode over an oxide cathode is the ability to operate the plate at higher voltages.

Figure 6-12 Photo of a typical oxide cathode. *(Courtesy of Varian Associates.)*

Oxide cathodes are susceptible to deterioration caused by ion bombardment. To achieve reasonable life, plate voltage must be limited. A thoriated tungsten cathode is more tolerant of ion bombardment, and so higher plate voltages can be safely applied.

The end of useful life for a thoriated tungsten tube occurs when most of the carbon has evaporated or has combined with residual gas, depleting the carbide surface layer. Theoretically, a 3 percent increase in filament voltage will result in a 20° K increase in cathode temperature, a 20 percent increase in peak emission, and a 50 percent decrease in tube life because of carbon loss. This cycle works in reverse, too. For a small decrease in temperature and peak emission, the life of the carbide layer — and hence, the tube — may be increased.

Tungsten-Impregnated Cathode. The tungsten-impregnated cathode typically operates at 900-1,175°C, and provides the highest average emission density of the three types of cathodes discussed here (500 mA/cm^2 to 8 A/cm^2). Peak current performance ranges up to 12 A/cm^2 or more.

Cathode Construction. Power tube filaments can be assembled in several different configurations. Figure 6-13 shows a spiral-type filament, and Figure 6-14 shows a bar-type design. The spiral filament is used extensively in low-power tubes. As the size of the tube

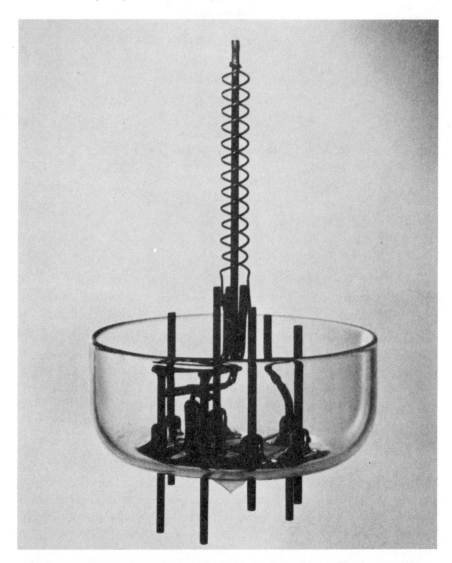

Figure 6-13 Spiral-type tungsten filament. *(Courtesy of Varian Associates.)*

increases, mechanical considerations dictate a bar-type filament with spring loading to compensate for thermal expansion. A mesh filament can be used for both small and large tubes. It is more rugged than other designs and less subject to damage from shock and vibration. The rigidity of a cylindrical mesh cathode depends on its diameter, on the number, thickness and length of the wires forming it, and on the fraction of welded to total wire crossings. A mesh cathode is shown in Figure 6-15.

Figure 6-14 Bar-type tungsten filament. *(Courtesy of Varian Associates.)*

Some power tubes are designed as a series of electron gun structures arranged in a cylinder around a center line. This construction
allows large amounts of plate current to flow and to be controlled
with a minimum amount of grid interception. With reduced grid in-

Figure 6-15 Mesh tungsten filament. *(Courtesy of Varian Associates.)*

terception, less power is dissipated in the grid structures. In the case of the control grid, less driving power is required for the tube.

In certain applications, the construction of the filament assembly can have an effect on the performance of the tube itself, and the performance of the RF system as a whole. For example, filaments built in a basket-weave mesh arrangement usually offer lower distortion in critical high-level AM modulation circuits.

6.2.3 Grid Structures

The type of grid used for a power tube is determined principally by the power level and operating frequency required. For most medium power tubes (5–25 kW dissipation), welded wire construction is common. At higher power levels, laser-cut *pyrolytic graphite* grids may be found. The grid structures of a power tube must maintain their shape and spacing at elevated temperatures. They must also withstand shock and vibration.

Wire Grids. Conventional wire grids are prepared by operators that wind the assemblies using special *mandrels* (forms) that include the required outline of the finished grid. The operators spot weld the wires at intersecting points (see Figure 6-16). Most grids of this type are made with tungsten or *molybdenum*, which exhibit stable physical properties at elevated temperatures. Grids for high power tubes are typically built with a bar-cage type of construction. A number of vertical supports are fastened to a metal ring at the top, and to a

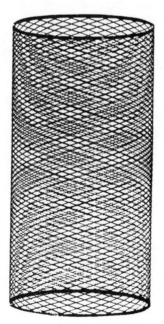

Figure 6-16 Mesh grid structure. *(Courtesy of Varian Associates.)*

base cone at the bottom. The lower end of the assembly is bonded to a contact ring. The construction of the ring, metal base cone, and cylindrical metal base give the assembly low lead inductance and low RF resistance.

The external loading of a grid during operation and the proximity of the grid to the hot cathode impose severe demands on both the mechanical stability of the structure and the physical characteristics of its surface. The grid absorbs a high proportion of the heat radiated by the cathode. It also intercepts the electron beam, converting part of the beam's kinetic energy into heat. Further, high frequency capacitive currents flowing in the grid result in additional heat. The end result is that grids are forced to work at temperatures as high as 1,500°C. Their primary and secondary emission must, however, be low. To prevent grid emission, high electron affinity must be ensured throughout the life of the tube, even though it is impossible to prevent material evaporated from the cathode from contaminating the grid surface. In tubes with oxide cathodes, grids made of tungsten or molybdenum wire are coated with gold to reduce primary emission caused by deposition. The maximum safe operating temperature for gold plating, however, is limited (about 550°C). Special coatings have, therefore, been developed for high temperature applications that are effective in reducing grid emission.

In tubes with thoriated tungsten cathodes, grids made of tungsten or molybdenum are coated with proprietary compounds to reduce primary emission. Primary grid emission is usually low in a thoriated tungsten cathode tube. In the case of an oxide cathode, however, free barium can evaporate from the cathode coating material and find its way to the control and screen grids. The rate of evaporation is a function of cathode temperature and time. A grid contaminated with barium will become another emitting surface. The hotter the grid, the greater the emissions.

K-grid. To permit operation at higher powers (and, therefore, higher temperatures) the so-called *K-grid* has been developed (Philips). The K-grid is a spot-welded structure of molybdenum wire doped to prevent brittleness and recrystallization. To eliminate mechanical stresses, the grid is annealed at 1,800°K. It is then baked in a vacuum at 2,300°K to de-gas the molybdenum wire. The grid structure is next coated with zirconium carbide (sintered at 2,300°K), and finally, with a thin layer of platinum. Despite high grid temperatures, the platinum coating keeps primary emissions to a minimum, while the surface structure keeps secondary emissions low.

Pyrolytic Grid. Pyrolytic grids are a high performance alternative to wire or bar grid assemblies. Used primarily for high power devices, pyrolytic grids are formed by laser-cutting a graphite cup of the proper dimensions. The computer-controlled laser cuts numerous holes in the cup to simulate a conventional-style grid. Figure 6-17 shows a typical pyrolytic-type grid before and after laser processing.

Pyrolytic (or oriented) graphite is a form of crystallized carbon produced by the decomposition of a hydrocarbon gas at very high temperatures in a controlled environment. A layer of pyrolytic graphite is deposited on a special form. The thickness of the layer is proportional to the amount of time deposition is allowed to continue. The structure and mechanical properties of the deposited graphite depend upon the imposed conditions.

Pyrolytic grids are ideal vacuum-tube elements because they do not expand like metal. Their small coefficient of expansion prevents movement of the grids inside the tube at elevated temperatures. This preserves the desired electrical characteristics of the device. Because tighter tolerances can be maintained, pyrolytic grids can be spaced more closely than conventional wire grids. Additional benefit inlcude:

• The grid is a single structure having no weld points.
• The grid has a thermal conductivity in two of the three planes nearly that of copper.
• It can operate at high temperatures with low vapor pressure.

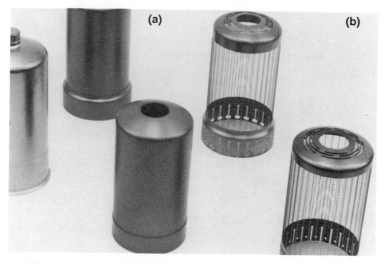

Figure 6-17 Pyrolytic graphite grid: (left) before laser processing; (right) completed assembly. *(Courtesy of Varian Associates.)*

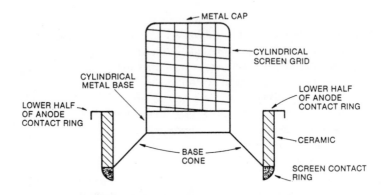

Figure 6-18 The screen grid assembly of a typical tetrode PA tube. (*Source: Broadcast Engineering Magazine.*)

Physical Structure. The control, screen and suppressor grids are cylindrical and concentric. Each is slightly larger than the previous grid, as viewed from the cathode. Each is fastened to a metal base cone, the lower end of which is bonded to a contact ring. Figure 6-18 shows the construction of a typical screen grid assembly. Figure 6-19 illustrates a cutaway view of a tetrode power tube. The shape of the control grid and its spacing from the cathode defines, in large part, the operating characteristics of the tube. For best performance, the

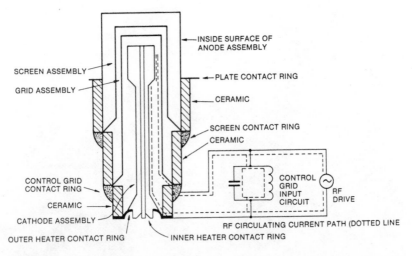

Figure 6-19 The internal arrangement of the anode, screen, control grid, and cathode assemblies of a tetrode power tube. (*Source: Broadcast Engineering Magazine.*)

grid must essentially be transparent to the electron path from the cathode to the plate. In a tetrode, the control and screen grids must be precisely aligned to minimize grid current. For pentode tubes, the previous two conventions apply, in addition to the requirement of precise alignment and minimum beam interception for the suppressor grid.

6.2.4   Plate Assembly

On the outside, a tube looks like one big unit, but it is really a collection of many smaller parts that must be machined and assembled to tight specifications. The anode and cooling fins (in the case of an air-cooled device) begin as flat sheets of copper. They are stamped by the tube manufacturer into the necessary sizes and shapes. Once all the parts have been machined, the anode and cooling fins are stacked in their proper positions, clamped, and brazed into one piece in a brazing furnace.

The plate of a power tube resembles a copper cup with the upper half of the contact ring welded to the mouth and the cooling fins silver-soldered or welded to the outside of the assembly. The lower half of the anode contact ring is bonded to a base ceramic spacer. At the time of assembly, the two halves of the ring are welded together to form a complete unit, as shown in Figure 6-20.

In most power tubes, the anode is a part of the envelope, and, because the outer surface is external to the vacuum, it can be cooled directly. Glass envelopes were used in older power tubes. Most have

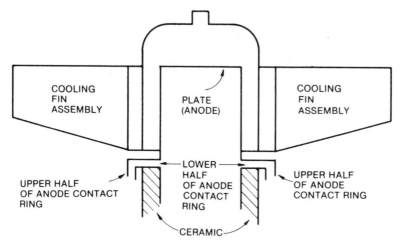

Figure 6-20 A cutaway view of the anode structure of an RF power amplifier tube. (*Source: Broadcast Engineering Magazine.*)

been replaced, however, with devices that use ceramic as the envelope material.

6.2.5   Tube Construction

Each type of power grid tube is unique insofar as its operating characteristics are concerned. The basic physical construction, however, is common to most devices. A vacuum tube is built in essentially two parts: the base, which includes filament and supporting stem, control grid, screen grid, and suppressor grid (if used); and the anode, which includes heat-dissipating fins made in various machining steps. The base subassembly is welded using a *tungsten-inert gas* (TIG) process in an oxygen-free atmosphere (a process sometimes referred to as *Heliarc* welding) to produce a finished base unit.

The ceramic elements used in a vacuum tube are critical to the device. Assembled in sections, each element builds upon the previous one to form the base of the tube. The ceramic-to-metal seals are created using a special material that is *painted* onto the ceramic and then heated in a brazing oven. After preparation in a high-temperature oven, the painted area provides a metallic structure that is molecularly bonded to the ceramic and a surface suitable for brazing. This process requires temperature sequences that dictate completion of the highest-temperature stages first. As the assembly takes form, lower oven temperatures are used so that completed bonds will not be damaged. Despite all the advantages that make ceramics one of the best insulators for a tube envelope, their brittleness is a potential cause of failure. The smallest cracks in the ceramic, not in themselves disastrous, can cause the ceramic to break when mechanically stressed by temperature changes.

After the base assembly has been matched with the anode, the completed tube is brazed into a single unit. The device then goes through a *bake-out* procedure. Baking stations are used to evacuate the tube and bake out any oxygen or other gases from the copper components of the assembly. Although oxygen-free copper is used in tube construction, some residual oxygen exists in the metal and must be driven out for long component life. A typical vacuum reading of 10^{-8} Torr is specified for most power tubes (formerly expressed as 10^{-8} mm of mercury). For comparison, this is the degree of vacuum in outer space about 200 mi. above the Earth.

To maintain a high vacuum during the life of the component, power tubes contain a *getter* device. The name comes from the function of the element: to "get" or trap and hold gases that may evolve inside the tube. Materials used for getters include zirconium, cerium, barium, and titanium.

The operation of a vacuum tube is an evolving chemical process. End of life is generally caused by loss of emission.

6.2.6 Connection Points

The high power levels and frequencies at which vacuum tubes operate place stringent demands on the connectors used to tie the outside world to the inside elements. Figure 6-21 shows a cutaway view of the base of a tetrode. Tubes are designed to be mounted vertically on their electrical connectors. The connectors provide a broad contact surface and mechanical support for the device. The cathode and grids are mounted on ring-shaped *Kovar* bases, which also serve as contact rings for the external connections. (Kovar is an iron-nickel-cobalt alloy whose coefficient of thermal expansion is comparable with that

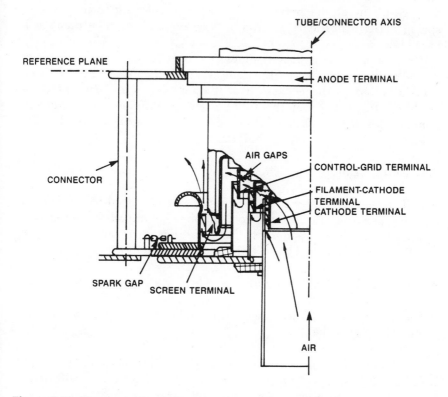

Figure 6-21 Cross-section of the base of a tetrode showing the connection points.

of aluminum oxide ceramic.) The different diameters of the various contact rings allow them to be grouped coaxially. The concentric tube/connector design provides for operation at high frequencies. Conductivity is improved by silver plating.

6.2.7 Tube Sockets

Any one tube design may have several possible socket configurations, depending upon the frequency of operation. If the tube terminals are large cylindrical surfaces, the contacting portions of the socket consist of either spring *collets* or an assembly of spring *finger-stock*. Usually, these multiple-contacting surfaces are made of beryllium copper to preserve spring tension at the high temperatures present at tube terminals. The fingers are silver-plated to reduce RF resistance. If the connecting fingers of a power-tube socket fail to provide adequate contact with the tube element rings, a concentration of RF currents will result. Depending upon the extent of this concentration, damage to the socket may result. Once a connecting finger loses its spring action, the heating effect is aggravated and tube damage is possible.

A series of tubes is also available with no sockets at all. Intended for cathode-driven service, the grid assembly is formed into a flange that is bolted to the chassis. The filament leads are connected via studs on the bottom of the tube. Such a configuration completely eliminates the tube socket. This type of device is useful for low-frequency applications, such as induction heating.

6.3 Neutralization

An RF power amplifier must be properly neutralized to provide acceptable performance in most applications. The means to accomplish this end can vary considerably from one design to another. An RF amplifier is neutralized when two conditions are met:

- The interelectrode capacitance between the input and output circuits is canceled.
- The inductance of the screen grid and cathode assemblies (in a tetrode) is canceled.

Cancellation of these common forms of coupling between the input and output circuits of vacuum tube amplifiers prevents self-oscillation and the generation of spurious products.

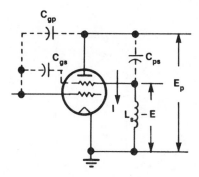

Figure 6-22 The elements involved in the neutralization of a tetrode PA stage. (*Source: Broadcast Engineering Magazine.*)

6.3.1 Neutralization Circuit Analysis

Figure 6-22 illustrates the primary elements that effect neutralization of a vacuum tube RF amplifier operating in the VHF band. (Many of the principles also apply to lower frequencies.) The feedback elements include the residual *grid-to-plate capacitance* (C_{gp}), *plate-to-screen capacitance* (C_{ps}), and screen grid lead inductance (L_s). The RF energy developed in the *plate circuit* (E_p) causes a current (I) to flow through the plate-to-screen capacitance and the screen lead inductance. The current through the screen inductance develops a voltage (-E) with a polarity opposite that of the plate voltage (E_p). The -E potential is often used as a method of neutralizing tetrode and pentode tubes operating in the VHF band.

Figure 6-23 graphically illustrates the electrical properties at work. The circuit elements of the previous figure have been arranged so that the height above or below the zero potential line represents magnitude and polarity of the RF voltage for each part of the circuit with respect to ground (zero). For the purposes of this illustration, assume that all of the circuit elements involved are pure reactances. The voltages represented by each, therefore, are either in phase or out of phase and can be treated as positive or negative with respect to each other. The voltages plotted in the figure represent those generated as a result of the RF output circuit voltage, E_p. No attempt is made to illustrate the typical driving current on the grid of the tube. The plate (P) has a high positive potential above the zero line, established at the ground point. Keep in mind that the distance above the

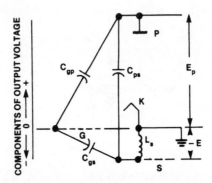

Figure 6-23 A graphic representation of the elements involved in the neutralization of a tetrode RF stage when self-neutralized. (*Source: Broadcast Engineering Magazine.*)

baseline represents increasing positive potential. The effect of the out-of-phase screen potential developed as a result of inductance L_s is shown, resulting in the generation of -E. As depicted, the figure constitutes a perfectly neutralized circuit. The grid potential rests at the zero baseline. The grid operates at filament potential insofar as any action of the output circuit on the input circuit is concerned.

The total RF voltage between plate and screen is comprised of the plate potential and screen lead inductance voltage, i.e., -E. This total voltage is applied across a divider circuit that is made up of the grid-to-plate capacitance and *grid-to-screen capacitance* (C_{gp} and C_{gs}). When this potential divider is properly matched for the values of plate RF voltage (E_p) and screen lead inductance voltage (-E), the control grid will exhibit zero voltage difference with respect to the filament as a result of E_p.

6.3.2   Neutralization Circuit Design

There are a variety of methods that may be used to neutralize a vacuum tube amplifier. Generally speaking, a grounded-grid, cathode-driven triode can be operated up to the VHF band without external neutralization components. The grounded-grid element is sufficient to prevent spurious oscillations. Tetrode amplifiers generally will operate through the MF band without neutralization. However, as the gain of the stage increases, the need to cancel feedback voltages caused by tube interelectrode capacitances and external connection inductances becomes more important. At VHF frequencies and above, it is generally necessary to provide some form of stage neutralization.

Below VHF. For operation at frequencies below the VHF band, neutralization for a tetrode typically employs a capacitance bridge circuit that will balance out the RF feedback caused by residual plate-to-grid capacitance. This method assumes that the screen is well bypassed to ground, providing the expected screening action inside the tube.

Neutralization of low-power push-pull tetrode or pentode tubes can be accomplished with cross-neutralization of the devices, as shown in Figure 6-24. Small value neutralization capacitors are used. In some cases, neutralization can be accomplished with a simple wire connected to each side of the grid circuit and brought through the chassis deck. Each wire is positioned to "look at" the plate of the tube on the opposite half of the circuit. Typically, the wire (or a short rod) is spaced 1 to 1/2 in. from the plate of each tube. Fine adjustment is accomplished by moving the wire in or out from its respective tube.

A similar method of neutralization can be used for a cathode-driven symmetrical stage (see Figure 6-25). Note that the neutralization capacitors, C_n, are connected from the cathode of one tube to the plate of the opposite tube. The neutralizing capacitors have a value equal to the internal cathode-to-plate capacitance of the PA tubes.

In the case of a single-ended amplifier, neutralization can be accomplished using either a push-pull output or push-pull input circuit.

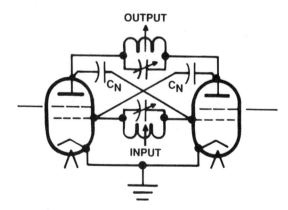

Figure 6-24 Push-pull grid neutralization. (*Source: Broadcast Engineering Magazine.*)

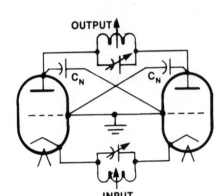

Figure 6-25 Symmetrical stage neutralization for a grounded-grid circuit. (*Source: Broadcast Engineering Magazine.*)

Figure 6-26 shows a basic push-pull grid neutralization scheme that provides the out-of-phase voltage necessary for proper neutralization. It is usually simpler to create a push-pull network in the grid circuit than the plate because of the lower voltages present. The *neutralizing capacitor*, C_n, is small and may consist of a simple feed-through wire (described previously). *Padding capacitor*, C_p, is often added to maintain the balance of the input circuit while tuning. C_p is generally equal in size to the input capacitance of the tube.

Single-ended tetrode and pentode stages can be neutralized using the method shown in Figure 6-27. The input resonant circuit is

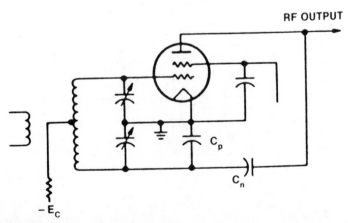

Figure 6-26 Push-pull grid neutralization in a single-ended tetrode stage. (*Source: Broadcast Engineering Magazine.*)

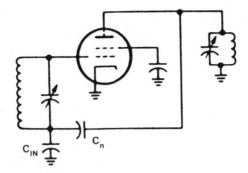

Figure 6-27 Single-ended grid neutralization for a tetrode. (*Source: Broadcast Engineering Magazine.*)

placed above ground by a small amount because of the addition of capacitor C_{in}. The voltage to ground that develops across C_{in} upon the application of RF drive is out of phase with the grid voltage, and is fed back to the plate through C_n to provide neutralization. In such a design, C_n is considerably larger in value than the grid-to-plate interelectrode capacitance.

The single-ended grid neutralization circuit is redrawn in Figure 6-28 to show the capacitance bridge that makes the design work. Balance is obtained when the following condition is met:

$$\frac{C_n}{C_{in}} = \frac{C_{gp}}{C_{gf}}$$

Where:
C_n = the neutralization capacitance
C_{in} = input circuit bypass capacitor
C_{gp} = grid-to-plate interelectrode capacitance
C_{gf} = total input capacitance, including tube and stray capacitance

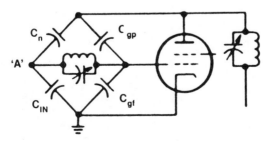

Figure 6-28 The previous figure redrawn to show the elements involved in neutralization. (*Source: Broadcast Engineering Magazine.*)

A single-ended amplifier can also be neutralized by taking the plate circuit slightly above ground and using the tube capacitances as part of the neutralizing bridge. This circuit differs from the usual RF amplifier design in that the plate bypass capacitor is returned to the screen side of the screen bypass capacitor (Figure 6-29). It should be noted that in the examples given, it is assumed that the frequency of operation is low enough so that inductances in the socket and connecting leads can be ignored. This is basically true in MF applications. At higher bands, however, the effects of stray inductances must be considered, especially in single-ended tetrode and pentode stages.

VHF And Above. Neutralization of power-grid tubes operating at VHF frequencies provides special challenges and opportunities to the design engineer. At VHF frequencies and above, significant RF voltages can develop in the residual inductances of the screen, grid, and cathode elements. If managed properly, these inductances can be used to accomplish neutralization in a simple, straightforward manner.

At VHF and above, neutralization is required to make the tube input and output circuits independent of each other with respect to reactive currents. Isolation is necessary to ensure independent tuning of the input and output. If variations in the output voltage of the stage produce variations of phase angle of the input impedance, phase modulation will result.

As noted previously, a circuit exhibiting independence between the input and output circuits is only half of the equation required for proper operation at RF frequencies. The effects of incidental inductance of the control grid must also be canceled for complete stability. This condition is required because the suppression of coupling by ca-

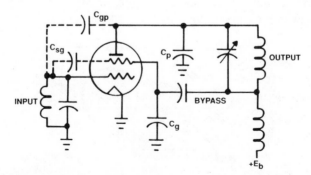

Figure 6-29 Single-ended plate neutralization. (*Source: Broadcast Engineering Magazine.*)

pacitive currents between the input and output circuits is not, by itself, sufficient to negate the effects of the output signal on the cathode-to-grid circuit. Both conditions, input and output circuit independence and compensation for control grid lead inductance, must be met for complete stage stability at VHF and above.

Figure 6-30 shows a PA stage employing stray inductance of the screen grid to achieve neutralization. In this grounded-screen application, the screen is tied to the cavity deck using six short connecting straps. Two additional adjustable ground straps are set to achieve neutralization.

Triode amplifiers operating in a grounded-grid configuration offer an interesting alternative to the more common grounded-cathode system. When the control grid is operated at ground potential, it serves to shield capacitive currents from the output to the input circuit. Typically, provisions for neutralization are not required until the point at which grid lead inductance becomes significant. This point is determined by the frequency of operation and the mechanical construction of the stage.

Two methods of neutralization are commonly used in grounded-grid RF amplifiers. The first requires an inductance between the grid and ground, or between the grids of a push-pull amplifier. This inductance can be used to compensate for the internal capacitance of the tube(s) and cancel coupling between the input and output cir-

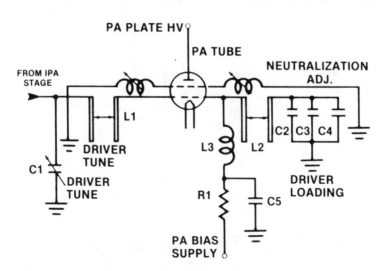

Figure 6-30 A grounded-screen PA stage neutralized through the use of stray inductance between the screen connector and the cavity deck. (*Source: Broadcast Engineering Magazine.*)

cuits. The second method involves connecting the grids of a push-pull amplifier to a point having zero impedance to ground. A bridge circuit of neutralizing capacitances is then added that is equal to the plate-to-filament capacitances of the tubes, thus achieving neutralization. These two neutralization schemes behave quite differently. They are special forms of neutralization in which the neutralizing capacitors have values differing from the internal capacitances of the tubes, and in which the appropriate reactance is inserted between the control grids.

6.3.3　Self-Neutralizing Frequency

The voltage dividing action between the *plate-to-grid capacitance* (C_{pg}) and the *grid-to-screen capacitance* (C_{gs}) will not change with changes in operating frequency. The voltage division between the plate and screen, and screen and ground caused by the charging current (I) will, however, vary significantly with frequency. There will be a particular frequency, therefore, where this potential dividing circuit will effectively place the grid at filament potential insofar as the plate is concerned. This point is known as the *self-neutralizing frequency*, illustrated in Figure 6-31.

At the self-neutralizing frequency, the tetrode or pentode is inherently neutralized by the circuit elements within the device itself, and the external screen inductance to ground. When a device is operated

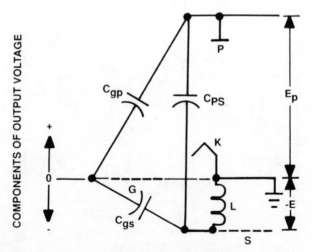

Figure 6-31 Graphical representation of the elements of a tetrode when self-neutralized. (Courtesy of Varian Associates.)

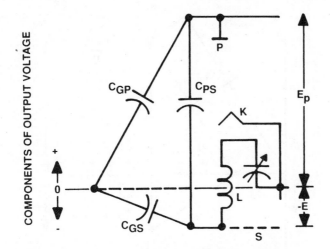

Figure 6-32 Components of the output voltage of a tetrode when neutralized by added series screen-lead capacitance. (*Courtesy of Varian Associates.)*

below its self-neutralizing frequency, the normal cross-neutralization circuits apply. When the operating frequency is above the self-neutralizing frequency, the voltage developed in the screen lead inductance is too large to give the proper voltage division between the internal capacitances of the device. One approach to neutralization in this case involves adjusting the inductive reactance of the screen lead to ground so as to lower the total reactance. In the example shown in Figure 6-32, this is accomplished with a series variable capacitor.

Another approach is shown in Figure 6-33, in which the potential divider network made up of the tube capacitance is changed. In the example, additional plate-to-grid capacitance is added external to the tube. The external capacitance (C_{ext}) can take the form of a small wire or rod positioned adjacent to the plate of the tube. This approach is similar to the one described previously for conventional neutralization, except that in this case the neutralizing probe is connected to the grid of the tube, rather than to an opposite polarity in the circuit.

Bibliography

1. Fink, D. and D. Christiansen, *Electronics Engineer's Handbook*, Third Edition, McGraw-Hill, New York, 1989.

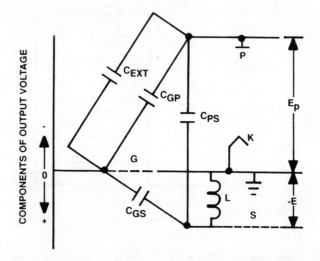

Figure 6-33 Components of the output voltage of a tetrode when neutralized by added external grid-to-plate capacitance. *(Courtesy of Varian Associates.)*

2. Jordan, Edward C., *Reference Data for Engineers: Radio, Electronics, Computer, and Communications,* Seventh Edition, Howard W. Sams Company, Indianapolis, IN, 1985.

3. *High Power Transmitting Tubes for Broadcasting and Research,* Philips Technical Publication, Eindhoven, The Netherlands, 1988.

7

Applying Vacuum Tubes

7.1 Introduction

Any number of configurations may be used to generate RF signals using vacuum tubes. Circuit design is dictated primarily by the operating frequency, output power, type of modulation, duty cycle, and available power supply. Tube circuits can be divided generally by their operating class and type of modulation employed.

7.1.1 Class C Amplification

The grounded cathode class C amplifier is the building block of RF technology. It is the simplest method of amplifying continuous wave (CW), pulsed, and FM signals. The basic configuration is shown in Figure 7-1. Tuned input and output circuits are used for impedance matching and to resonate the stage at the desired operating frequency. The cathode is bypassed to ground using low value capacitors. Bias is applied to the grid as shown. The bias power supply may be eliminated if a self-bias configuration is used.

The typical operating efficiency of a class C stage ranges from 65 to 85 percent. Depending on the degree of linearity required from the amplifier, class B operation may be used. Under such conditions, typical efficiency drops to 60 to 70 percent.

Grounded-Grid Amplifier. Figure 7-2 illustrates the application of a zero-bias triode in a grounded-grid arrangement. Because the grid operates at RF ground potential, this circuit offers stable opera-

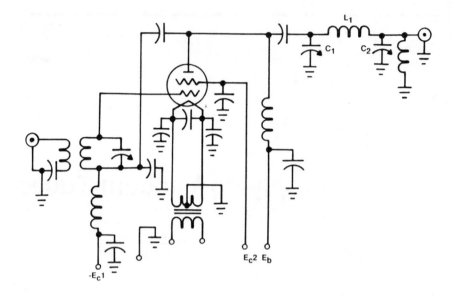

Figure 7-1 Basic grounded cathode class C RF amplifier circuit. *(Courtesy of Varian Associates.)*

tion without the need for neutralization (at MF and below). The input signal is coupled to the cathode through a matching network. The output of the triode feeds a pi network through a blocking capacitor.

7.1.2 Amplitude Modulation

Class C plate modulation is the classic method of producing an AM waveform. Figure 7-3 shows the basic circuit. Triodes and tetrodes may be used as the modulator or carrier tube. Triodes offer the simplest and most common configuration.

Numerous variations exist on the basic design, including a combination of plate and screen grid modulation. The carrier signal is applied to the control grid and the modulating signal is applied to the screen and plate. The plate is fully modulated and the screen is modulated 70 to 100 percent to achieve 100 percent carrier modulation. Modulation of the screen can be accomplished using one of the following methods:

- Screen voltage supplied through a dropping resistor connected to the unmodulated dc plate supply, as shown in Figure 7-4(a).
- An additional (third) winding on the modulation transformer, illustrated in Figure 7-4(b).

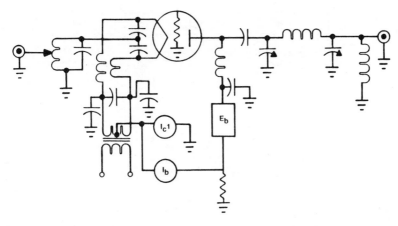

Figure 7-2 Typical amplifier circuit using a zero-bias triode. Grid current is measured in the return lead from ground to the filament. *(Courtesy of Varian Associates.)*

• A modulation choke placed in series with a low voltage fixed screen supply, shown in Figure 7-4(c).

Depending on the design, screen/plate modulation may also require partial modulation of the control grid to achieve the desired performance characteristics.

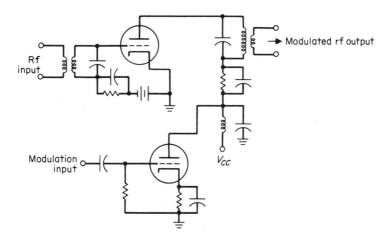

Figure 7-3 The classic plate-modulated PA circuit for AM applications. (*Source:* Fink and Christiansen, *Electronics Engineers' Handbook, 3rd Ed.,* McGraw-Hill, New York, 1989.)

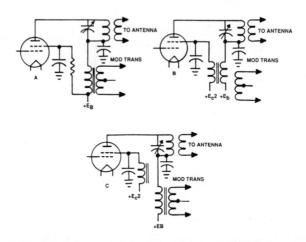

Figure 7-4 Basic methods of screen and plate modulation. (a) plate modulation. (b) plate and screen modulation. (c) plate modulation. (Courtesy of Varian Associates.)

Control Grid Modulation. Although not as common as class C plate modulation, a class C RF amplifier may be modulated by varying the voltage on the control grid of a triode. This approach, shown in Figure 7-5, produces a change in the magnitude of the plate current pulses and, therefore, variations in the output waveform taken from the plate tank. Both the carrier and modulation signals are applied to the grid.

The primary benefit of grid modulation is obvious: a modulator is not required. Grid modulation, however, requires a fixed plate supply

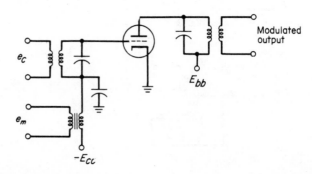

Figure 7-5 Grid modulation of a class C RF amplifier. (Source: Fink and Christiansen, Electronics Engineers' Handbook, 3rd Ed., McGraw-Hill, New York, 1989.)

voltage that is twice the peak RF voltage without modulation. The result is higher plate dissipation at lower modulation levels, including carrier level. The typical plate efficiency of a grid-modulated stage may range from only 35 to 45 percent at carrier. Grid modulation is typically used only in systems where plate modulation transformers cannot provide adequate bandwidth for the intended application.

Suppressor Grid Modulation. The output of a class C pentode amplifier may be controlled by applying a modulating voltage, superimposed on a suitable bias, to the suppressor grid in order to produce an AM waveform. As the suppressor grid becomes more negative, the minimum instantaneous potential at which current can be drawn to the plate is increased. Thus, as modulation varies the suppressor-grid potential, the output voltage changes. This method of modulating an RF amplifier provides about the same plate efficiency as the grid modulated stage. Overall operating efficiency, however, is slightly lower because of increased screen grid losses associated with the design. Linearity of the circuit is not usually high.

Cathode Modulation. Cathode modulation incorporates the principles of both control grid and plate modulation. As shown in Figure 7-6, a modulation transformer in the cathode circuit varies the grid-cathode potential, as well as the plate-cathode potential. The ratio of grid modulation to plate modulation is set by adjusting the tap shown in the figure. Grid leak bias is typically employed to improve linearity of the stage.

7.2 High Efficiency Linear Amplification[1]

The class B RF amplifier is the classical means of achieving linear amplification of an AM signal. A significant efficiency penalty, however, is paid for linear operation. The plate efficiency at carrier level in a practical linear amplifier is about 33 percent, while a high-level (plate) modulated stage can deliver an efficiency of 65 to 80 percent. In a real-world transmitter, power consumed by the modulator stage must also be taken into account. However, because the output power

---

1 Portions of Section 7.2 were adapted from *Electronics Engineer's Handbook*, Third Edition, Fink and Christiansen editors, (Chapter 13, "High Power Amplifiers") McGraw-Hill, 1989.

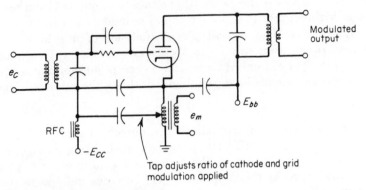

Top adjusts ratio of cathode and grid
modulation applied

Figure 7-6 Cathode-modulation of a class C RF amplifier. (*Source:* Fink and Christiansen, *Electronics Engineers' Handbook, 3rd Ed.*, McGraw-Hill, New York, 1989.)

of the modulator is at most one-third of the total power output, the net efficiency of the plate modulated amplifier is still higher than a conventional linear amplifier of comparable power.

Still, linear amplification is attractive to the users of high power transmitters because low-level modulation can be employed. This eliminates the need for modulation transformers and reactors, which — at certain power levels and for particular applications — may be difficult to design and expensive to construct. To overcome the efficiency penalty of linear amplification, special systems have been designed that take advantage of the benefits of a linear mode, while not imposing an excessive efficiency burden. Linear amplifiers, in fact, find numerous applications in high power transmission systems.

7.2.1 Chireix Outphasing Modulated Amplifier

The *Chireix* amplifier employs two output stage tubes. The grids of the tubes are driven with signals whose relative phase varies with the applied modulating signal. The output waveforms of the tubes are then applied to a summing network, and finally to the load. Figure 7-7 shows a simplified schematic diagram of the Chireix system.

Operation of the circuit is straightforward. When the grids of the tubes are driven with signals that are 180° out of phase, power output will be zero. This corresponds to a negative AM modulation peak of 100 percent. Full (100 percent) positive modulation occurs when both grids are driven with signals that are in-phase. The AM modulating signal is produced by varying the relative phase of the two drive signals.

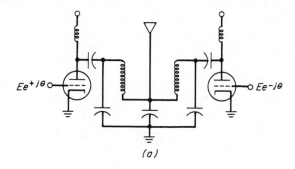

(a)

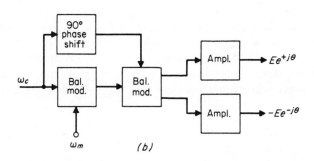

(b)

Figure 7-7 Simplified schematic diagram of the Chirex outphasing modulated amplifier: (a) output circuit; (b) drive signal generation circuit. (*Source:* Fink and Christiansen, *Electronics Engineers' Handbook, 3rd Ed,* McGraw-Hill, New York, 1989.)

The phase-modulated carrier frequency grid waveforms are generated in a low-level balanced modulator and amplified to the level required by the PA tubes. Stability of the exciter in the Chireix system is critical. Any shift in the relative phase of the exciter outputs will translate to amplitude modulation of the transmitted waveform.

Because of the outphasing method employed to produce the AM signal, the power factor associated with the output stage is of special significance. The basic modulating scheme would produce a unity power factor at zero output. The power factor would then decrease with increasing RF output. To avoid this undesirable characteristic, the output plate circuits are detuned, one above resonance and one below, to produce a specified offset. This *dephasing*, coupled with adjustment of the relative phase of the driving signals, keeps the power factor within a reasonable range. The linearity of the Chireix amplifier is good. Typical overall efficiency is 60 percent.

7.2.2   Doherty Amplifier

The *Doherty* modulated amplifier is probably the most common high efficiency system used in high power AM transmitters. The system is a true high efficiency linear amplifier, rather than a hybrid amplifier/modulator. The Doherty amplifier employs two tubes and a 90° network as an *impedance inverter* to achieve load-line modification as a function of power level. The basic circuit is shown in Figure 7-8.

Two tubes are used in the Doherty amplifier, a *carrier tube* (V1) and a *peak tube* (V2). The carrier tube is biased class B. Loading and drive for the tube are set to provide maximum linear voltage swing at carrier level. The peak tube is biased class C; at carrier it just begins to conduct plate current. Each tube delivers an output power equal to twice the carrier power when working into a load impedance of R Ω. At carrier, the reflected load impedance at the plate of V1 is 2R Ω. This is the correct value of load impedance for carrier-level output at full plate voltage swing. (The impedance-inverting property

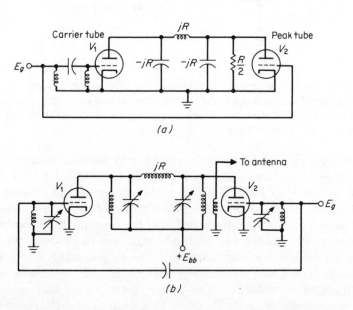

(a)

(b)

Figure 7-8 Doherty high-efficiency AM amplifier: (a) operating theory; (b) schematic diagram of a typical system. (*Source:* Fink and Christiansen, *Electronics Engineers' Handbook, 3rd Ed.,* McGraw-Hill, New York, 1989.)

of the 90° network is similar to a 1/4-wave transmission line.) A 90° phase lead network is included in the grid circuit of V1 to compensate for the 90° phase lag produced by the impedance inverting network in the plate circuit. For negative modulation swings, the carrier tube performs as a linear amplifier with a load impedance of 2R Ω; the peak tube contributes no power to the output.

On positive modulation swings, the peak tube conducts and contributes power to the R/2 load resistance. This is equivalent to connecting a negative resistance in shunt with the R/2 load resistance, so that the value seen at the load end of the 90° network increases. This increase is reflected through the network as a decrease in load resistance at the plate of V1, causing an increase in output current from the tube, and hence an increase in output power. The drive levels on the tubes are adjusted so that each contributes the same power (equal to twice the carrier power) at a positive modulation peak. For this condition, a load of R Ω is presented to each tube.

The key aspect of the Doherty circuit is the change in load impedance on tube V1 with modulation. This property enables the device to deliver increased output power at a constant plate voltage swing. The end result is high efficiency and good linearity. Overall efficiency of the Doherty amplifier ranges from 60 to 65 percent. A practical application of the Doherty circuit is shown in Figure 7-8(b). The shunt reactances of the phase-shift networks are achieved by detuning the related tuned circuits. The tuned circuits in the grid are tuned above the operating frequency; those in the plate are tuned below the operating frequency.

Screen Modulated Doherty-type Amplifier. A variation on the basic Doherty scheme can be found in the screen modulated (Continental Electronics) power amplifier. The design is unique in that the system does not have to function as a linear amplifier. Modulation is, instead, applied to the screen grids of both the carrier and peak tubes. The peak tube is modulated upward during the positive half of the modulation cycle, and the carrier tube is modulated downward during the negative half of the cycle.

A screen modulated amplifier designed for MF operation at 1 MW is shown in Figure 7-9. The carrier tube delivers all of the 1 MW carrier power to the load through the 90° impedance inverting network. When modulation is applied to the screen of the carrier tube (during the negative half-cycle), the carrier tube output follows the modulation linearly. Because the carrier tube is driven to full swing, its voltage excursion does not increase during the positive half-cycle.

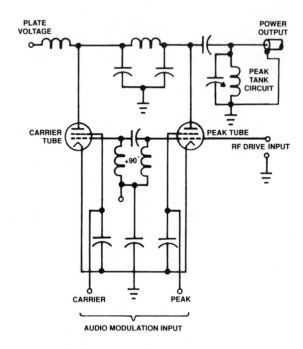

Figure 7-9 Screen-modulated Doherty-type circuit. *(Courtesy of Varian Associates.)*

Positive modulation is needed on the screen only to maintain the full plate swing as the impedance load changes on the carrier tube.

The peak tube is normally operated with a negative screen voltage that maintains plate current at near cutoff. No power, therefore, is delivered until modulation is applied. During the positive peak of modulation, the screen of the peak tube is modulated upward. As the peak tube is modulated positive, it delivers power into the output circuit. As the power delivered increases, the load seen by the inter-plate network changes as in the Doherty system.

7.2.3   Terman-Woodyard Modulated Amplifier

The *Terman-Woodyard* configuration uses the basic scheme of Doherty for achieving high efficiency (the impedance-inverting property of a 90° phase-shift network). However, the Terman-Woodyard design also employs grid modulation of both the carrier tube and the peak tube. This allows both tubes to be operated class C. The result is an increase in efficiency over the Doherty configuration. The basic

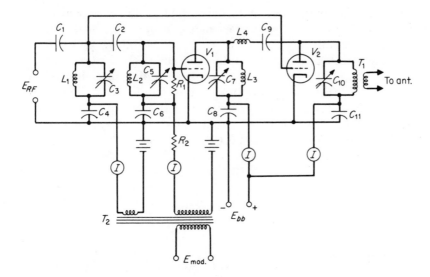

Figure 7-10 Terman-Woodyard high-efficiency modulated amplifier.
(*Source:* Fink and Christiansen, *Electronics Engineers' Handbook, 3rd Ed.,* McGraw-Hill, New York, 1989.)

circuit is shown in Figure 7-10. With no modulation, V1 operates as a class C amplifier, supplying the carrier power, and V2 is biased so that it is just beginning to conduct. The efficiency at carrier is, therefore, essentially that of a class C amplifier.

During positive modulation swings, V2 conducts. At 100 percent modulation, both tubes supply equal amounts of power to the load, similar to the Doherty design. During negative modulation swings, V2 is cut off and V1 performs as a standard grid-modulated amplifier.

The Terman-Woodyard system rates efficiency as a function of modulation percentage for sinusoidal waveforms. Typical efficiency at carrier is 80 percent, falling to a minimum of 68 percent at 50 percent modulation and 73 percent at 100 percent modulation.

7.2.4 Dome Modulated Amplifier

The *Dome* high efficiency amplifier employs three power output tubes driven by different audio signals. Modulation is achieved by load-line modification during positive modulation excursions, and by linear amplification during negative modulation swings. Load-line modification is achieved by absorption of a portion of the generated

RF power. However, most of the absorbed power is returned to the plate power supply, rather than being dissipated as heat. High efficiency is the result.

A basic Dome modulated amplifier is shown in Figure 7-11. Tube V1 is used in a plate-modulated configuration, supplying power to the grid of V2 (the power amplifier tube). The load impedance in the plate circuit of V2 is equivalent to the load impedance reflected into the primary of transformer T1 (from the antenna) in series with the impedance appearing across the C8 terminals of the 90° phase-shift network consisting of C8, C9, and L4. The impedance appearing across the C8 terminals of the 90° network is inversely related to the impedance across the other terminals of the network (in other words, the effective ac impedance of V3). Thus, with V3 cut off, a short circuit is reflected across the C8 terminals of the network. Tube V3 performs as a modulated rectifier with audio supplied to its grid. Dome calls this tube the *modifier*.

With no modulation (the carrier condition), tube V3 is biased to cutoff, and the drive to V2 is adjusted until power output is equal to four times the carrier power (corresponding to a positive modulation peak). Note that all the plate signal voltage of V2 appears across the primary of transformer T1 for this condition. The bias on V3 is then reduced, lowering the ac impedance of the tube and, therefore, reflecting an increasing impedance at the C8 terminals of the 90° net-

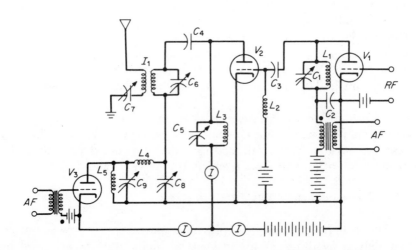

Figure 7-11 Dome high-efficiency modulated amplifier. (*Source:* Fink and Christiansen, *Electronics Engineers' Handbook, 3rd Ed.,* McGraw-Hill, New York, 1989.)

work. The bias on V3 is reduced until the operating carrier power is delivered to the antenna. An amount of power equal to the carrier is thus rectified by V3 and returned to the plate power supply, except for that portion dissipated on the V3 plate. The drive level to V2 is finally adjusted so that the tube is just out of saturation.

For positive modulation excursions, V3 grid voltage is driven negative, reaching cutoff for a positive modulation peak. For negative modulation swings, V2 acts as a linear amplifier. V3 does not conduct during the negative modulation swing because the peak RF voltage on the plate is less than the dc supply voltage on the cathode.

Because of the serial loss incurred in the power amplifier tube and in the modifier tube for energy returned to the power supply, the Dome circuit is not as efficient as the Doherty configuration. Typical efficiency at carrier ranges from 50 to 60 percent.

7.3 Special Application Amplifiers

The operating environment of tubes used in industrial applications is usually characterized by widely varying supply voltages, heavy and variable vibration, and significant changes in loading. RF systems designed for industrial use typically involve induction (eddy-current) heating of materials and dielectric heating of plastics. The relative drive power required and the efficiency of the RF generator is of less importance than the stability of the output stage. It is not unusual for a maximum four percent variation in output power to be permitted with load variations of 40 percent on either side of the nominal value.

Scientific applications are no less demanding. Research projects span a wide range of powers and frequencies. Typical uses for tetrodes as of this writing include:

- Super proton synchrotron delivering 2.4 MW at 200 MHz
- Fusion reactor utilizing 33 MW of high frequency heating in the 25-60 MHz band
- Plasma research using 12 MW in the 60-120 MHz range
- Fusion research using a 500 kW ion cyclotron resonance heating generator operating at 30-80 MHz

It can be seen from these examples that heat dissipation is a major consideration in the design of an amplifier or oscillator for scientific use.

7.3.1   Distributed Amplification

Specialized research and industrial applications often require an amplifier covering several octaves of bandwidth. At microwave frequencies, a TWT may be used. At MF and below, multiple tetrodes are usually combined to achieve the required bandwidth. A *distributed amplifier* consists of multiple tetrodes and lumped-constant transmission lines. The lines are terminated by load resistances with magnitudes equal to their characteristic impedances. A cyclical wave of current is present on the output transmission line circuit, with each tube contributing its part in proper phase. A typical configuration for a distributed amplifier includes 8 to 16 tubes. The efficiency of such a design is low.

7.4   Cavity Amplifier Systems

Power grid tubes are ideally suited for use as the power generating element in a cavity amplifier. Because of the physical dimensions involved, cavity designs are typically limited to VHF frequencies and above. Lower frequency RF generators utilize discrete L and C devices to create the needed tank circuit. Cavity amplifiers are viewed by many equipment users as just so much black magic. The theory of operation, however, is fairly straightforward, and like most things in electronics, cavity amplifiers are easy to understand, once you understand them. Two types of cavity amplifiers are commonly used: 1/2-wavelength and 1/4-wavelength systems.

7.4.1   Current Paths

The operation of a cavity amplifier is an extension of the current paths inside the tube. Two elements must be examined in this discussion: the input circulating currents, and the output circulating currents.

Input Circuit. The grid/cathode assembly resembles a transmission line whose termination is the RF resistance of the electron stream within the tube. Figure 7-12 shows the current path of an RF generator (the RF driver stage output) feeding a signal into the grid/cathode circuit. The outer contact ring of the cathode heater assembly makes up the inner conductor of a transmission line, formed by the cathode and control grid assemblies. The filament wires are returned down the center of the cathode. For the input circuit to work correctly, the cathode must have a low RF impedance to

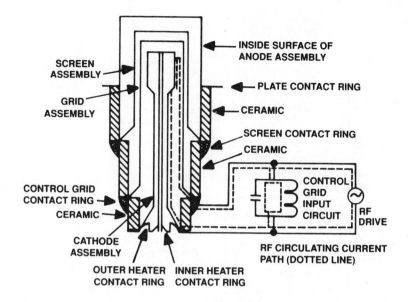

Figure 7-12 A simplified representation of the grid input circuit of a PA tube. (*Source: Broadcast Engineering Magazine.*)

ground. This cathode bypassing may be accomplished in one of several ways.

Below 30 MHz, the cathode can be grounded to RF voltages by simply bypassing the filament connections with capacitors, as shown in Figure 7-13(a). Above 30 MHz, this technique does not work well because of the stray inductance of the filament leads. Notice that in Figure 7-13(b), the filament leads appear as RF chokes, preventing the cathode from being placed at RF ground potential. This causes negative feedback and reduces the efficiency of the input and output circuits. In Figure 7-13(c) the cathode circuit is configured to simulate a 1/2-wave transmission line. The line is by-passed to ground with large value capacitors 1/2-wavelength from the center of the filament (at the filament voltage feed point). This transmission line RF short is repeated 1/2-wavelength away at the cathode (heater assembly) and effectively places it at ground potential.

Because 1/2-wavelength bypassing is usually bulky at VHF frequencies (and may be expensive), RF generators are often designed using certain values of inductance and capacitance in the filament/cathode circuit that will create an artificial transmission line which simulates a 1/2-wavelength shorted transmission line. As illustrated in Figure 7-13(c), the inductance and capacitance of the fila-

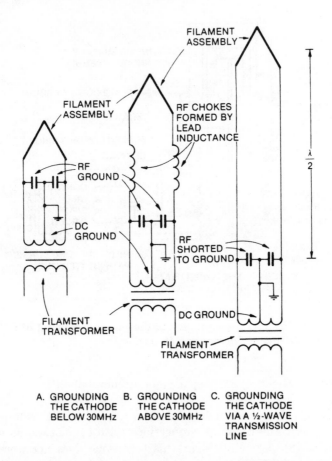

A. GROUNDING   B. GROUNDING   C. GROUNDING
THE CATHODE   THE CATHODE   THE CATHODE
BELOW 30MHz   ABOVE 30MHz   VIA A ½-WAVE
                                             TRANSMISSION
                                             LINE

Figure 7-13 Three common methods for RF bypassing of the cathode of a tetrode PA tube: (a) grounding the cathode below 30 MHz; (b) grounding the cathode above 30 MHz; (c) grounding the cathode via a 1/2-wave transmission line. (*Source: Broadcast Engineering Magazine.*)

ment circuit can resemble an artificial transmission line of 1/2-wavelength if the values of L and C are properly selected.

Output Circuit. The plate-to-screen circulating current of the tetrode is shown in Figure 7-14. For the purposes of example, consider that the output RF current is generated by an imaginary current generator located between the plate and screen grid. The RF current travels along the inside surface of the plate structure (because of the skin effect), through the ceramic at the lower half of the anode contact ring, across the bottom of the fins, and to the band around the

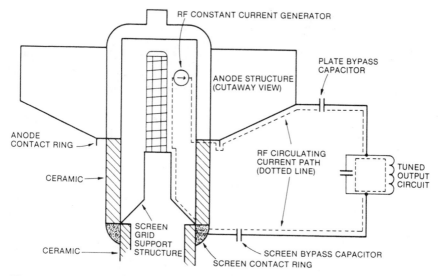

Figure 7-14 RF circulating current path between the plate and screen in a tetrode PA tube. (*Source: Broadcast Engineering Magazine.*)

outside of the fins. The RF current then flows through the plate by-pass capacitor to the RF tuned circuit and load, and returns to the screen grid. The return current travels through the screen bypass capacitor (if used) and screen contact ring, up the screen base cone to the screen grid, and back to the imaginary generator.

The screen grid has RF current returning to it, but because of the assembly's low impedance, the screen grid is effectively at RF ground potential. The RF current generator, therefore, appears to be feeding an open-ended transmission line consisting of the anode (plate) assembly and the screen assembly. The RF voltage developed by the anode is determined by the plate impedance (Z_p) presented to the anode by the resonant circuit and its load.

When current flows on one conductor of a transmission line cavity circuit, an equal magnitude current flows in the opposite direction on the other conductor. This means that a large value of RF-circulating current is flowing in a cavity amplifier outer conductor (the cavity box). All of the outer conductor circulating currents start at and return to the screen grid (in a tetrode-based system). The front or back access panel (door) of the cavity is part of the outer conductor and large values of circulating current flow into it, through it, and out of it. A mesh contact strap is generally used to electrically connect the access panel to the rest of the cavity.

7.4.2   1/4-Wavelength Cavity

The 1/4-wavelength PA cavity is common in transmitting equipment. The design is simple and straightforward. A number of variations can be found in different RF generators, but the underlying theory of operation is the same.

A typical 1/4-wave cavity is shown in Figure 7-15. The plate of the tube connects directly to the inner section (tube) of the plate-blocking capacitor. The exhaust chimney/inner conductor forms the other plate of the blocking capacitor. The cavity walls form the outer conductor of the 1/4-wave transmission line circuit. The dc-plate voltage is applied to the PA tube by a cable routed inside the exhaust chimney and inner tube conductor. In the design shown in the figure, the screen-contact fingerstock ring mounts on a metal plate that is insulated from the grounded-cavity deck by a sheet of insulating material. This hardware makes up the screen-blocking capacitor assembly. The dc-screen voltage feeds to the fingerstock ring from underneath the cavity deck through an insulated feedthrough assembly.

A grounded-screen configuration may also be used in this design in which the screen-contact fingerstock ring is connected directly to the grounded cavity deck. The PA cathode then operates at below ground potential (in other words, at a negative voltage), establishing the required screen voltage to the tube.

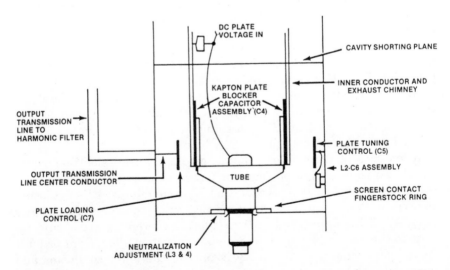

Figure 7-15 The physical layout of a common type of 1/4-wavelength PA cavity designed for VHF operation. (*Source: Broadcast Engineering Magazine.*)

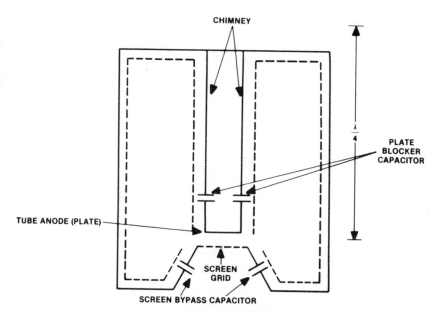

Figure 7-16 The RF circulating current paths for the 1/4-wavelength cavity shown the previous figure. (*Source: Broadcast Engineering Magazine.*)

The cavity design shown in the figure is set up to be slightly shorter than a full 1/4-wavelength at the operating frequency. This makes the load inductive and resonates the tube's output capacity. Thus, the physically foreshortened shorted transmission line is resonated and electrically lengthened to 1/4-wavelength.

Figure 7-16 illustrates the paths taken by the RF-circulating currents in the circuit. RF energy flows from the plate, through the plate-blocking capacitor, along the inside surface of the chimney/inner conductor (because of the skin effect), across the top of the cavity, down the inside surface of the cavity box, across the cavity deck, through the screen-blocking capacitor, over the screen-contact fingerstock, and into the screen grid.

Figure 7-17 shows a graph of RF current, voltage, and impedance for a 1/4-wavelength coaxial transmission line. It shows that infinite impedance, zero RF current, and maximum RF voltage occur at the feed point. This would not be suitable for a practical PA circuit because arcing would result from the high RF voltage, and poor efficiency would be caused by the mismatch between the tube and the load. Notice, however, the point on the graph marked at slightly less than 1/4-wavelength. This length yields an impedance of 600-800 Ω

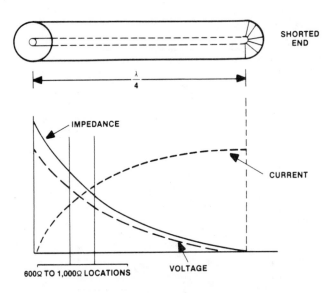

Figure 7-17 Graph of the RF current (...), RF voltage (- - -), and RF impedance (___) for a 1/4-wavelength shorted transmission line. At the feed point RF current is zero, RF voltage is maximum, and RF impedance is infinite. (*Source: Broadcast Engineering Magazine.*)

and is ideal for the PA-plate circuit. It is necessary, therefore, to physically foreshorten the shorted coaxial transmission-line cavity to provide the correct plate impedance. Shortening the line also is a requirement for resonating the tube's output capacity, because the capacity shunts the transmission line and electrically lengthens it.

Figure 7-18 shows a graph of the RF current, voltage, and impedance presented to the plate of the tube as a result of the physically foreshortened line. This plate impedance is now closer to the ideal 600-800 Ω value required by the tube anode.

Tuning the cavity. Coarse tuning of the cavity is accomplished by adjusting the cavity length. The top of the cavity (the cavity shorting deck) is fastened by screws or clamps and can be raised or lowered to set the length of the cavity for the particular operating frequency. Fine tuning is accomplished by a variable-capacity plate-tuning control built into the cavity. In a typical design, one plate of the tuning capacitor — the stationary plate — is fastened to the inner conductor just above the plate-blocking capacitor. The movable tuning plate is fastened to the cavity box, the outer conductor, and is mechanically linked to the front-panel tuning control. This capacity shunts the inner conductor to the outer conductor and is used to vary the electrical length and resonant frequency of the cavity.

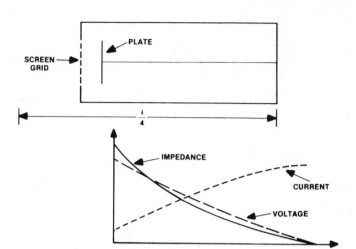

Figure 7-18 Graph of the RF current (...), RF voltage (- - -), and RF impedance (___) produced by the physically foreshortened coaxial transmission line cavity. (*Source: Broadcast Engineering Magazine.*)

The 1/4-wavelength cavity is inductively coupled to the output port. This coupling is usually on the side opposite the cavity access door. The inductive pickup loop can take several forms. In one design it consists of a half-loop of flat copper bar stock that terminates in the loading capacitor at one end and feeds the output transmission line inner conductor at the other end (Figure 7-19). The inductive pickup ideally would be placed at the maximum current point in the 1/4-wavelength cavity. However, this point is located at the cavity shorting deck, and when the deck is moved for coarse tuning the magnetic coupling will be changed. A compromise in positioning, therefore, must be made. The use of a broad, flat copper bar for the coupling loop adds some capacitive coupling to augment the reduced magnetic coupling.

Adjustment of the loading capacitor couples the 50 Ω transmission-line impedance to the impedance of the cavity. Heavy loading lowers the plate impedance presented to the tube by the cavity. Light loading reflects a higher load impedance to the amplifier plate.

7.4.3 1/2-Wavelength Cavity

The operation of a 1/2-wavelength cavity follows the same basic reasoning as the 1/4-wavelength design outlined previously. The actual construction of the system will depend upon the power level used and the required bandwidth. An example will help to illustrate how a 1/2-wavelength cavity operates.

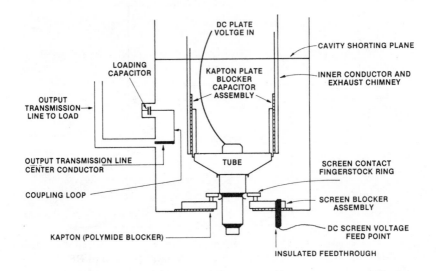

Figure 7-19 Basic loading mechanism for a VHF 1/4-wave cavity. (*Source: Broadcast Engineering Magazine.*)

The design of a basic 1/2-wavelength PA cavity for operation in the region of 100 MHz (the FM broadcast band) is shown in Figure 7-20. The tube anode and a silver-plated brass pipe serve as the inner conductor of the 1/2-wave transmission line. The cavity box serves as the outer conductor. The transmission line is open at the far end and repeats this condition at the plate of the tube. The line is, in effect, a parallel resonant circuit for the PA tube.

The physical height of the circuit shown (67 in.) was calculated for operation down to 88 MHz. To allow adequate clearance at the top of the transmission line and space for input circuitry at the bottom of the assembly, the complete cavity box would have to be almost 8 ft. tall. This is too large for any practical transmitter.

Figure 7-21 shows RF voltage, current, and impedance for the inner conductor of the transmission line and the anode of the tube. The load impedance at the plate is thousands of ohms. The RF current is, therefore, extremely small and the RF voltage is extremely large. In the application of such a circuit, arcing would become a problem. The high plate impedance would also make amplifier operation inefficient. The figure also shows an area between the anode and the 1/4-wavelength location where the impedance of the circuit is 600-800 Ω. As noted previously, this value is ideal for the anode of the PA tube. To achieve this plate impedance, the inner conductor

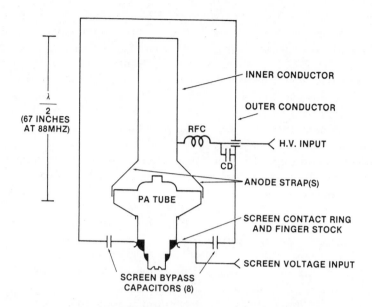

Figure 7-20 The 1/2-wavelength PA cavity in its basic form. (*Source: Broadcast Engineering Magazine.*)

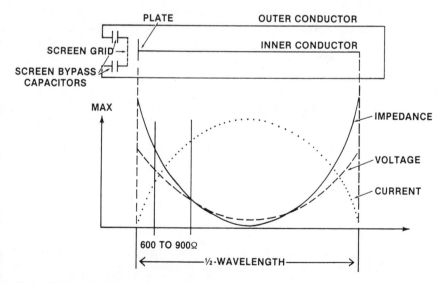

Figure 7-21 The distribution of RF current (...), voltage (- - -), and impedance (___) along the inner conductor of a 1/2-wavelength cavity. (*Source: Broadcast Engineering Magazine.*)

must be less than a full 1/2-wavelength. The physically foreshortened transmission-line circuit must, however, be electrically resonated (lengthened) to 1/2-wavelength for proper operation. If the line length were changed to operate at a different frequency, the plate impedance also would change because of the new distribution of RF voltage and current on the new length of line. The problem of frequency change, therefore, is twofold: the length of the line must be adjusted for resonance, and the plate impedance of the tube must be kept constant for good efficiency.

To accommodate operation of this system at different frequencies, while keeping the plate impedance constant, two forms of coarse tuning and one form of fine tuning are built into the 1/2-wave PA cavity. Figure 7-22 shows the tube and its plate line (inner conductor). The inner conductor is U-shaped to reduce the cavity height. With the movable section (the plate tune control) fully extended, the inner conductor measures 38 in. and the anode strap measures 7 in. The RF path from the anode strap to the inside of the tube plate (along the surface because of the skin effect) is estimated to be about 8 in. This makes the inner conductor maximum length about 53 in. This is too short to be a 1/2-wavelength at any FM frequency. The full length of a 1/2-wave line is 54.7 in. at 108 MHz and 67.1 in. at 88 MHz.

The coarse-tuning and fine-tuning provisions of the cavity, coupled with the PA tube's output capacity, resonate the plate line to the

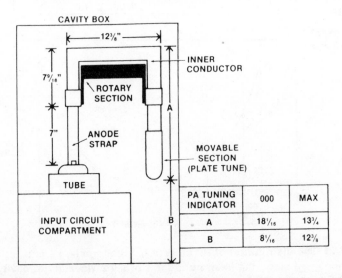

Figure 7-22 The configuration of a practical 1/2-wavelength PA cavity. (*Source: Broadcast Engineering Magazine.*)

exact operating frequency. In effect, they electrically lengthen the physically foreshortened line. This process, along with proper loading, determines the plate impedance and, therefore, the efficiency.

Lengthening the Plate Line. The output capacity of the tube is the first element that electrically lengthens the plate line. A 1/2-wave transmission line that is too short offers a high impedance that is both resistive and inductive. The output capacity of the tube resonates this inductance. The detrimental effects of the tube's output capacity are, therefore, eliminated. The anode strap and the cavity inner conductor rotary section provide two methods of coarse frequency adjustment for resonance.

The anode strap (shown in Figure 7-23) has less of a cross-sectional area than the inner conductor of the transmission line. It, therefore, has more inductance than an equal length of the inner conductor. The anode-coupling strap acts as a series inductance and electrically lengthens the plate circuit.

At low frequencies one narrow anode strap is used. At mid-FM frequencies one wide strap is used. The wide strap exhibits less inductance than the narrow strap and does not electrically lengthen the plate circuit as much. At the upper end of the FM band, two anode straps are used. The parallel arrangement lowers the total inductance of the strap connection and adds still less electrical length to the plate circuit.

The main section of the plate resonant line, together with the rotary section, function as a parallel inductance. RF current flows in the same direction through the transmission line and the rotary sec-

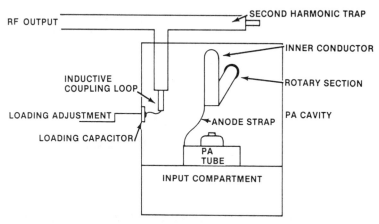

Figure 7-23 Coarse tuning mechanisms for the 1/2-wave cavity. (*Source: Broadcast Engineering Magazine.*)

tion. Therefore, the magnetic fields of the two paths are combined. When the rotary section is at maximum height, the magnetic coupling between the main section of the transmission line and the rotary assembly is maximum. Because of the relatively large mutual inductance provided by this close coupling, the total inductance of these parallel inductors increases. This electrically lengthens the transmission line and lowers the resonant frequency. The concept is illustrated in Figure 7-24(a).

When the rotary section is at minimum height, the magnetic coupling between the two parts of the inner conductor is minimum. This reduced coupling lowers the mutual inductance, which lowers the total inductance of the parallel combination. The reduced inductance allows operation at a higher resonant frequency. This condition is illustrated in (b). The rotary section provides an infinite number of coarse settings for various operating frequencies.

The movable plate-tune assembly is located at the end of the inner-plate transmission line. It can be moved up and down to change the physical length of the inner conductor by about 4 3/4 in. This assembly is linked to the front-panel plate-tuning knob, providing a fine adjustment for cavity resonance.

7.4.4 Folded 1/2-Wave Cavity

A special case of the 1/2-wavelength PA cavity is shown in Figure 7-25. The design employs a folded 1/2-wave resonator constructed with coaxial aluminum and copper tubing. This cavity arrangement eliminates the high-voltage blocking capacitor and high-current shorting contacts of conventional designs by connecting the main transmission-line resonant circuit conductor directly to the anode of the power tube. A grounded, concentric transmission-line center con-

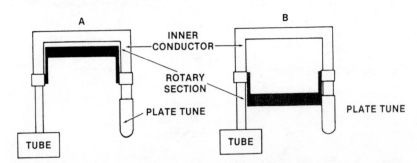

Figure 7-24 Using the cavity's movable section to adjust for resonance: (a) the rotary section at maximum height; (b) the rotary section at minimum height. (*Source: Broadcast Engineering Magazine.*)

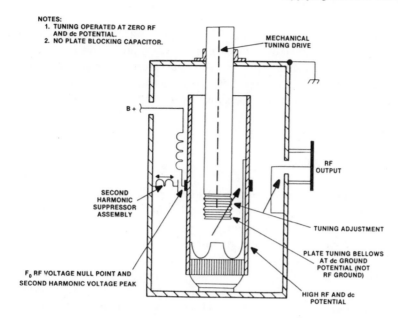

NOTES:
1. TUNING OPERATED AT ZERO RF AND dc POTENTIAL.
2. NO PLATE BLOCKING CAPACITOR.

MECHANICAL TUNING DRIVE

B +

SECOND HARMONIC SUPPRESSOR ASSEMBLY

RF OUTPUT

TUNING ADJUSTMENT

PLATE TUNING BELLOWS AT dc GROUND POTENTIAL (NOT RF GROUND)

F_0 RF VOLTAGE NULL POINT AND SECOND HARMONIC VOLTAGE PEAK

HIGH RF AND dc POTENTIAL

Figure 7-25 The basic design of a folded 1/2-wavelength cavity. (*Source: Broadcast Engineering Magazine.*)

ductor tunes the cavity with a variable re-entrant length inserted into the end of the main conductor opposite the tube.

The main conductor (the fixed portion of the plate line) is insulated from the ground and carries the anode dc potential. For easy RF decoupling, high-voltage power is fed at the fundamental frequency RF-voltage null point, approximately 1/4-wavelength from the anode. A large surface area without sliding contacts results in minimal loss.

Incorporated into the tank design is a second-harmonic suppressor. Rather than attenuating the second harmonic after the signal has been generated and amplified, this design essentially prevents formation of second-harmonic energy by series-LC trapping the second-harmonic waveform at the point where the wave exhibits a high impedance (approximately 1/4-wavelength from the anode). The second harmonic will peak in voltage at the same point that the dc-plate potential is applied.

Plate tuning is accomplished by an adjustable bellow on the center portion of the plate line, which is maintained at chassis ground potential. Output coupling is accomplished with an untuned loop intercepting the magnetic field concentration at the voltage null (maximum RF current) point of the main line. The PA-loading control varies the angular position of the plane of the loop with respect to the plate line, changing the amount of magnetic field that it intercepts. Multiple phosphor-bronze leaves connect one side of the output loop to ground and the other side to the center conductor of the output transmission line. This allows for mechanical movement of the loop by the PA-loading control without using sliding contacts. The grounded loop improves immunity to lightning and static build-up on the antenna.

7.4.5   Output Coupling

Coupling is the process by which RF energy is transferred from the amplifier cavity to the output transmission line. Wideband cavity systems use coupling to transfer energy from the primary cavity to the secondary cavity. Coupling in tube-type power amplifiers usually transforms a high (plate or cavity) impedance to a lower output (transmission line) impedance. Both capacitive (electrostatic) and inductive (magnetic) coupling methods are used in cavity RF amplifiers. In some designs, combinations of the two methods are used.

Inductive Coupling. Inductive coupling employs the principles of transformer action. The efficiency of the coupling depends upon three conditions:

* The cross-sectional area under the coupling loop, compared to the cross-sectional area of the cavity (see Figure 7-26). This effect can be compared to the turns ratio principle of a transformer.
* The orientation of the coupling loop to the axis of the magnetic field. Coupling from the cavity is proportional to the cosine of the angle at which the coupling loop is rotated away from the axis of the magnetic field, as illustrated in Figure 7-27.
* The amount of magnetic field that the coupling loop intercepts (see Figure 7-28). The strongest magnetic field will be found at the point of maximum RF current in the cavity. This is the area where maximum inductive coupling is obtained. Greater magnetic field strength also is found closer to the center conductor of the cavity. Coupling, therefore, is inversely proportional to the distance of the coupling loop from the center conductor.

In both 1/4- and 1/2-wavelength cavities, the coupling loop generally feeds a 50 Ω transmission line (the load). The loop is in series with the load and has considerable inductance at VHF frequencies. This inductance will reduce the RF current that flows into the load, thus reducing power output. This effect can be overcome by placing a

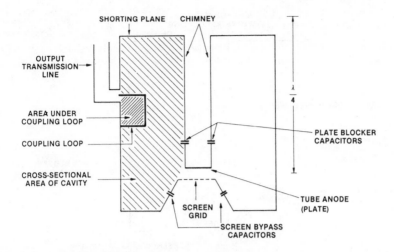

Figure 7-26 The use of inductive coupling in a 1/4-wavelength PA stage. (*Source: Broadcast Engineering Magazine.*)

variable capacitor in series with the output coupling loop. The load is connected to one end of the coupling loop and the variable capacitor ties the other end of the loop to ground. The variable capacitor cancels some or all of the loop inductance. It functions as the PA-stage loading control.

Maximum loop current and output power occurs when the loading capacitor cancels all of the inductance of the loading loop. This lowers the plate impedance and results in heavier loading. Light loading results if the loading capacitance does not cancel all of the loop inductance. The loop inductance that is not canceled causes a decrease in load current and power output, and an increase in plate impedance.

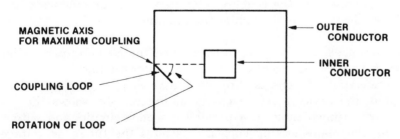

Figure 7-27 Top view of a cavity showing the inductive coupling loop rotated away from the axis of the magnetic field of the cavity. (*Source: Broadcast Engineering Magazine.*)

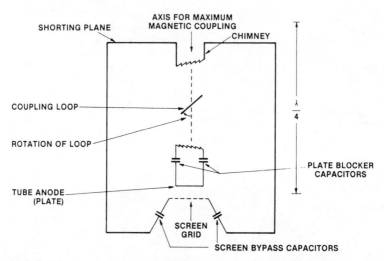

Figure 7-28 Cutaway view of the cavity showing the coupling loop rotated away from the axis of the magnetic field. (*Source: Broadcast Engineering Magazine.*)

Capacitive Coupling. Capacitive (electrostatic) coupling, which physically appears to be straightforward, often baffles the applications engineer because of its unique characteristics. Figure 7-29 shows a cavity amplifier with a capacitive-coupling plate positioned near its center conductor. This coupling plate is connected to the output load, which can be a transmission line or a secondary cavity (for wideband operation). The parameters that control the amount of capacitive coupling include:

• The area of the coupling-capacitor plate (the larger the area, the greater the coupling).
• The distance from the coupling plate to the center conductor (the greater the distance, the lighter the coupling).

Maximum capacitive coupling occurs when the coupling plate is at the maximum voltage point on the cavity center conductor.

To understand the effects of the capacitive coupling, the equivalent circuit of the cavity must be observed. Figure 7-30 shows the PA tube cavity (functioning as a parallel resonant circuit) and output section. The plate-blocking capacitor isolates the tube's dc voltage from the cavity. The coupling capacitor and output load are physically in series, but electrically they appear to be in parallel, as shown in Figure 7-31. The resistive component of the equivalent par-

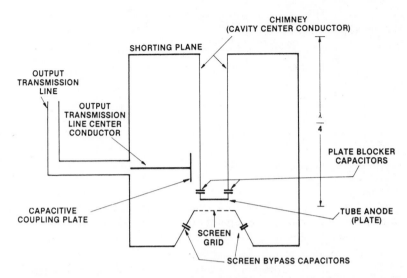

Figure 7-29 1/4-wavelength cavity with capacitive coupling to the output load. (*Source: Broadcast Engineering Magazine.*)

allel circuit is increased by the coupling reactance. The equivalent parallel coupling reactance is absorbed into the parallel resonant circuit. This explains the necessity to retune the plate after changing the PA stage coupling (loading). The coupling reactance can be a series capacitor or inductor.

The series-to-parallel transformations are explained by the following formula:

$$R_p = \frac{R_s^{\ 2} + X_s^{\ 2}}{R_s} \quad \text{and} \quad X_p = \frac{R_s^{\ 2} + X_s^{\ 2}}{X_s}$$

Figure 7-30 The equivalent circuit of a 1/4-wavelength cavity amplifier with capacitive coupling. (*Source: Broadcast Engineering Magazine.*)

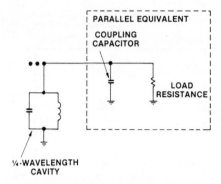

Figure 7-31 The equivalent circuit of a 1/4-wavelength cavity amplifier showing how series capacitive coupling appears electrically as a parallel circuit to the PA tube. (*Source: Broadcast Engineering Magazine.*)

Where:
R_p = effective parallel resistance
R_s = actual series resistance
X_s = actual series reactance
X_p = effective parallel reactance

PA Loading. Proper loading of a cavity PA stage to the output transmission line is critical to the dependable operation and optimum efficiency of the overall system. Light coupling produces light loading and results in a high plate impedance; conversely, heavy coupling results in heavier loading and a lower plate impedance. Maximum output power, coinciding with maximum efficiency and acceptable dissipation, dictates a specific plate impedance for a cavity of given design. This plate impedance is also dependent upon the values of dc plate voltage (E_p) and plate current (I_p).

Plate impedance dictates the cavity parameters of loaded Q, RF circulating current, and bandwidth. The relationships can be expressed as follows:

* Loaded Q is directly proportional to the plate impedance and controls the other two cavity parameters.

$$\text{Loaded } Q = \frac{Z_p}{X_l}$$

Where:
Z_p = cavity plate impedance
X_l = cavity inductive reactance

- Circulating current in the cavity is much greater (by a factor of the loaded Q) than the RF current supplied by the tube.
  Circulating current = Q × I$_p$

  Where:
  I$_p$ = the RF current supplied to the cavity by the tube

- The cavity bandwidth is dependent on the loaded Q and operating frequency.

  $$\text{Bandwidth} = \frac{F_r}{Q}$$

  Where:
  F$_r$ = the cavity resonant frequency

Heavy loading lowers the PA plate impedance and cavity Q. A lower Q reduces the cavity RF circulating currents. In some cavities, high circulating currents can cause cavity heating and premature failure of the plate or screen blocking capacitors. The effects of lower plate impedance — a byproduct of heavy loading — are higher RF and dc plate currents, and reduced RF plate voltage. Instantaneous plate voltage is the result of RF plate voltage added to dc plate voltage. The reduced swing of plate voltage causes a less positive dc screen current to flow. Positive screen current flows only when the plate voltage swings close to or below the value of the positive screen grid.

Light loading raises the plate impedance and cavity Q. A higher Q will increase the cavity circulating currents, raising the possibility of component overheating and failure. The effects of higher plate impedance are reduced RF and dc plate current and increased RF and dc plate voltage excursions. The higher cavity RF or peak dc voltages are, the greater the possibility that arcing will occur in the cavity. There is one value of plate impedance that will yield optimum output power, efficiency, dissipation, and dependent operation. It is dictated by cavity design, and the values of the various dc and RF voltages and currents supplied to the stage.

Depending on the cavity design, light loading may seem deceptively attractive. The dc plate voltage is constant (set by the power supply), and the lower dc plate current resulting from light loading reduces the tube's overall dc input power. The RF output power may change with light loading, depending on the plate impedance and cavity design, while efficiency will probably increase or, at worst, remain constant. Caution must be exercised with light loading, however, because of the increased RF voltages and circulating currents that such operation creates. The manufacturer's recommendations on PA tube loading should, therefore, be carefully observed.

Although there are many similarities among various cavity designs, each one imposes its own set of operational requirements and limitations. No two cavity systems will tune up in exactly the same fashion. Given proper maintenance, a cavity amplifier will provide years of reliable service.

7.4.6 Mechanical Design

Understanding the operation of a cavity amplifier is usually difficult because of the nature of the major component elements. The capacitors don't necessarily look like capacitors, and the inductors don't necessarily look like inductors. It is often difficult to relate the electrical schematic diagram to the mechanical assembly that exists within the transmitter. At VHF and UHF frequencies — the domain of cavity PA designs — inductors and capacitors can be formed out of some strange-looking mechanical devices and hardware.

Consider the PA cavity schematic diagram shown in Figure 7-32. The grounded-screen stage is of conventional design. Decoupling of the high-voltage power supply is accomplished by C1, C2, C3, and L1. Capacitor C3 is located inside the PA chimney (cavity inner conductor). The RF sample lines provide two low-power RF outputs for a modulation monitor or other test instrument. Neutralization inductors L3 and L4 consist of adjustable grounding bars on the screen grid ring assembly.

Figure 7-33 shows the electrical equivalent of the PA cavity schematic diagram. The 1/4-wavelength cavity acts as the resonant tank for the PA. Coarse tuning of the cavity is accomplished by adjustment of the shorting plane. Fine tuning is performed by the PA tuning control, which acts as a variable capacitor to bring the cavity into resonance. The PA loading control consists of a variable capacitor that couples the cavity to the load. The assembly made up of L2 and C6 prevents spurious oscillations within the cavity.

The logic of a PA stage often disappears when you are confronted with the actual physical design of the system. Blocking capacitor C4 is constructed of a roll of *Kapton* insulating material sandwiched between two circular sections of aluminum. (Kapton is a registered trademark of DuPont.) PA plate tuning control C5 consists of a large surface area aluminum plate that can be moved in or out of the cavity to reach resonance. PA loading control C7 is constructed much

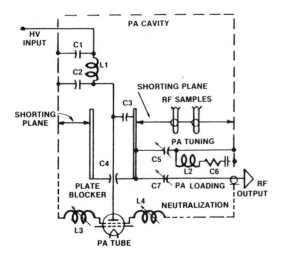

Figure 7-32 Typical VHF cavity amplifier. (*Source: Broadcast Engineering Magazine.*)

the same as the PA tuning assembly, with a large-area paddle feeding the harmonic filter, located externally to the cavity. The loading paddle may be moved toward the PA tube or away from it to achieve the required loading. The L2-C6 damper assembly actually consists of a 50 Ω noninductive resistor mounted on the side of the cavity wall. Component L2 is formed by the inductance of the connecting strap between the plate tuning paddle and the resistor. Component C6 is the equivalent stray capacitance between the resistor and the surrounding cavity box.

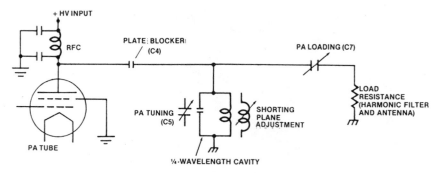

Figure 7-33 Electrical equivalent of the cavity amplifier shown in the previous figure. (*Source: Broadcast Engineering Magazine.*)

It can be seen that cavity amplifiers involve as much mechanical engineering as they do electrical engineering. Figure 7-34 shows graphically the level of complexity that a cavity amplifier may involve. The photo depicts a VHF power amplifier (Philips) with broadband input circuitry.

7.5   Vacuum Tube Life

Failures in semiconductor components result primarily from deterioration of the device caused by exposure to environmental fluctuations and voltage extremes. The vacuum tube, on the other hand, suffers wear out because of a predictable chemical reaction. Life expectancy is one of the most important factors to be considered in the use of vacuum tubes. While it is impossible to predict the life of an individual device, accurate estimates of average life expectancy can be determined for a group of tubes operating under similar conditions.

With the many possible variables in the operation of an RF generator (including filament voltage, ambient temperature, power output, and frequency of operation), it is difficult to say with any amount of accuracy what the *average* life for a given tube might be. On-site

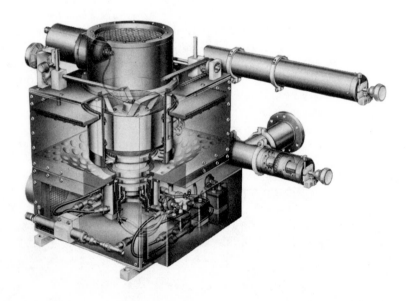

Figure 7-34 Cross-section view of a broadband VHF cavity amplifier.
(Courtesy of Philips.)

experience is the best indicator. Possible causes of short tube life include the following:

- Poor filament voltage regulation
- Insufficient cooling system air flow (or water flow)
- Improper transmitter tuning
- Inaccurate panel meters or external wattmeter, resulting in more demand from the tube than is actually required
- Improper stage neutralization

In general, manufacturers specify maximum operating parameters for power grid tubes so that operation within the ratings will provide for a minimum useful life of 1,000 hours. Tubes used in pulsed service must be considered separately.

7.5.1 Cathode Life

The cathode is the heart of any power tube. The device is said to wear out when filament emissions are inadequate for full power output or acceptable distortion levels. In the case of a thoriated tungsten filament tube, three primary factors determine the number of hours a device will operate before reaching this condition:

- The rate of evaporation of thorium from the cathode
- The quality of the tube vacuum
- The operating temperature of the filament

In the preparation of thoriated tungsten, one to two percent of thorium oxide (thoria) is added to the tungsten powder before it is sintered and drawn into wire form. After being mounted in the tube, the filament is usually carburized by being heated to a temperature of about 2,000° K in a low pressure atmosphere of hydrocarbon gas or vapor until its resistance increases by ten to 25 percent. This process allows the reduction of the thoria to metallic thorium. The life of the filament as an emitter is increased because the rate of evaporation of thorium from the carburized surface is several times smaller than from a surface of pure tungsten.

Despite the improved performance obtained by carburization of a thoriated-tungsten filament, they are susceptible to deactivation by the action of positive ions. Although the deactivation process is negligible for anode voltages below a critical value, a trace of residual gas pressure too small to affect the emission from a pure tungsten filament can cause rapid deactivation of a thoriated-tungsten filament.

This restriction places stringent requirements on vacuum processing the tube.

These factors taken together determine the wear-out rate of the tube. Catastrophic failures because of inter-electrode shorts or failure of the vacuum envelope are considered abnormal and are usually the result of some external influence. Catastrophic failures not the result of the operating environment are usually caused by a defect in the manufacturing process. Such failures generally occur early in the life of the component.

The design of the transmission system can have a substantial impact on the life expectancy of a vacuum tube. Protection circuitry must remove applied voltages rapidly to prevent damage to the tube in the event of a failure external to the device. Sufficient cooling air must be directed toward the base of the tube and the anode cooling assembly (or water in the case of a water or vapor cooled device) whenever high voltage is applied.

The filament turn-on circuit can also have an effect on PA tube life expectancy. The surge current of the filament circuit must be maintained at a low level to prevent thermal cycling problems. This consideration is particularly important in medium- and high-power PA tubes.

When the heater voltage is applied to a cage-type cathode, the tungsten wires expand immediately because of their low thermal inertia. However, the cathode support, which is made of massive parts (relative to the tungsten wires), expands more slowly. The resulting differential expansion can cause permanent damage to the cathode wires. It can also cause a modification of the tube characteristics, and occasionally arcs between the cathode and the control grid.

7.5.2   Filament Voltage

Long tube life requires filament voltage regulation. A tube whose filament voltage is allowed to vary along with the primary ac line voltage will not achieve the life expectancy possible with a tightly regulated supply. This problem is particularly acute at mountain-top installations, where utility regulation is generally poor.

To accurately adjust the filament voltage, a *true reading* RMS voltmeter is required. A true-reading RMS meter is preferred over the more common *average responding* RMS meter because the true-reading meter can accurately measure a voltage despite an input waveform that is not a pure sine wave (as would be the case in an SCR-regulated filament supply).

To extend tube life, some engineers in the communications and broadcast industries leave the filaments of power grid tubes on at all times, not shutting down at sign-off. If the sign-off period is three hours or less, this practice may be beneficial. Filament voltage regulation is a must in such situations because the primary line voltages may vary substantially from the carrier-on to carrier-off value.

Filament Voltage Management. The life of a thoriated tungsten power tube can often be extended considerably through accurate filament voltage management. It is not uncommon to double the useful life of the device. The procedure, recommended by some tube manufacturers, is as follows:

1. Operate the filament at its full-rated voltage for the first 200 hours following installation.
2. After this burn-in period, reduce the filament voltage in steps of about 0.1 V until power output begins to fall (for FM or CW systems) or until modulating waveform distortion begins to increase (for AM systems).
3. When the *emissions floor* has been reached, raise the filament voltage about 0.2 V. Long-term operation at this voltage can result in a substantial extension in the usable life of the tube, as illustrated in Figure 7-35.
4. At regular intervals, about every three months, check the filament voltage and increase it if power output begins to fall or distortion begins to rise.
5. For one week of each year of tube operation, run the filament at full-rated voltage. This will operate the getter and clean the tube of gas.
6. Never operate the tube with a filament voltage that is at or below 90 percent of its rated operating voltage. Never increase the filament voltage to more than 105 percent of normal.

Some tube manufacturers place the minimum operating point at 94 percent. Others recommend that the tube be set for 100 percent filament voltage and left there. The choice of which approach to take is left to the user.

Warm-up/Cool-down. Many tube manufacturers specify a filament warm-up period of several minutes before the application of plate voltage. This reduces the thermal stresses placed on the tube. Some RF generators include a time-delay relay to prevent the appli-

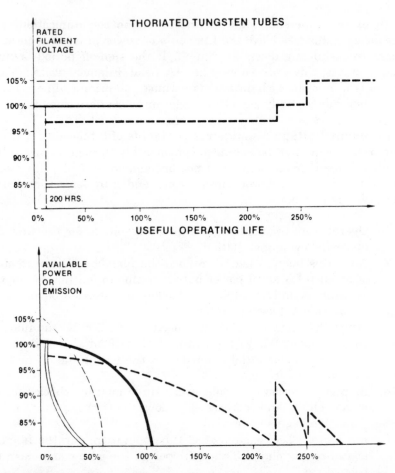

Figure 7-35 The effects of filament voltage management on the useful life of a thoriated tungsten filament power tube. Note the dramatic increase in emission hours when filament voltage management is practiced. (*Source: Broadcast Engineering Magazine.*)

cation of a plate-on command until a predetermined warm-up cycle is completed. Manufacturers may also specify a recommended cool-down period between the application of *plate-off* and *filament-off* commands. This cool-down, generally five to ten minutes, is designed to prevent excessive temperatures on the PA tube surface when the cooling air is shut off. Large vacuum tubes contain a significant mass of metal, which stores heat quite effectively. Unless cooling air is maintained at the base of the tube and through the anode cooling structure (or water in the case of a water cooled device), an excessive rise in temperature can occur. Again, the result can be shorter tube life, or even catastrophic failure because of seal cracks caused by thermal stress. Most tube manufacturers suggest that cooling air continue to be directed toward the tube base and anode after filament voltage has been removed to further cool the device.

7.5.3 Tuning Considerations

There are probably as many ways to tune the PA stage of an RF generator or transmitter as there are types of systems. Experience is the best teacher when it comes to adjusting for peak efficiency and performance. Compromises often must be made among various operating parameters. Tuning can be affected by any number of changes in the PA stage. Replacing the final tube in an AM broadcast transmitter or low- to medium-frequency RF generator usually does not significantly change the stage tuning. Replacing a tube in an FM or TV transmitter, or high frequency RF generator, can, on the other hand, significantly alter stage tuning. At high frequencies, normal tolerances and variations in tube construction result in changes in element capacitance and inductance that may affect operation of the system.

Stability is one of the primary objectives of transmitter tuning. Peak efficiency, performance, and stability, unfortunately, do not always coincide. Trade-offs must sometimes be made to ensure proper operation of the system.

Backheating. Backheating is an unusual, but potentially destructive, phenomenon that can affect tubes operating at VHF and above. At high frequencies, the transit time of electrons from the cathode through the grid to the plate becomes an appreciable part of the cycle. Electrons can, in fact, be stopped in transit and turned back by a rapidly changing grid voltage. In this situation, electrons are deflected from their normal path and given excess energy with which they bombard the cathode and other portions of the tube structure. The likelihood of this condition occurring is minimized through the use of the lowest possible RF grid voltage on the tube, obtained by using the lowest possible dc grid bias. Backheating is typically a problem only for triodes. Tetrodes usually do not exhibit this phenomenon because of the accelerating action of the screen grid. Further, because of the higher gain of a tetrode, lower grid voltages are used.

Loading. A power tube should always be operated with its specified load. If the plate load is removed so that the minimum instantaneous plate voltage tends to fall to values around cathode potential, the number of electrons tuned back can destroy the device. Experience has shown that operating a tube under no-load conditions can cause sufficient heating to crack the ceramic envelope.

Parasitic oscillations at VHF and above can lead to tube failure through the same mechanisms as light or no loading. Parasitic oscillations are seldom loaded heavily; the designer does not expect to experience them. If parasitic oscillations occur, a tube can be damaged or destroyed in the same manner as in unloaded operation, even though the transmitter is loaded properly for the fundamental operating frequency. If an unloaded VHF parasitic is present simultaneously with apparently satisfactory operation on the fundamental, short tube life may result.

7.5.4   Cooling System

The most critical points of almost every PA tube type are the metal-to-ceramic junctions or seals. At temperatures below 250°C these seals remain secure, but above that temperature the bonding in the seal may begin to disintegrate. Warping of grid structures also may occur at temperatures above the maximum operating level of the tube. The result of prolonged overheating is shortened tube life or catastrophic failure.

Air Cooling Considerations. All modern air-cooled PA tubes use an air system socket and matching chimney for cooling. Normally the tube socket is mounted in a pressurized compartment so that cooling air passes through the socket and then is guided to the anode cooling fins, as illustrated in Figure 7-36.

Providing sufficient cooling air at the socket assembly is important for proper cooling of the tube anode and base, and for cooling of the contact rings of the tube itself. The contact fingers used in the *collet* assembly of a socket typically are made of beryllium copper. If subjected to temperatures above 150°C for an extended length of time, the beryllium copper will lose its temper (springy characteristic) and will no longer make good contact with the base rings of the tube. In extreme cases, such socket problems can lead to arcing, which can burn through the metal portion of the base ring. Such an occurrence can ultimately lead to catastrophic failure of the device because of loss of the vacuum envelope.

Several precautions are usually taken to prevent damage to tube seals under normal operating conditions. Air directors or sections of tubing may be used to provide spot-cooling to critical surface areas of

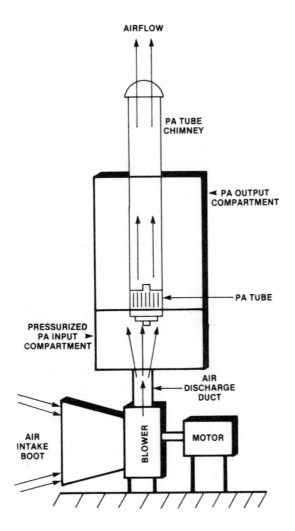

Figure 7-36 Airflow system for an air-cooled power tube. *(Source: Broadcast Engineering Magazine.)*

the device. Airflow sensors are generally installed at one or more key locations to prevent operation of the system in the event of a cooling system failure.

Water/Vapor Cooling System. Water purity is the key to long life in a water or vapor cooling system. Depending on the transmitter, the cooling system may be required to dissipate tens or even hundreds of kilowatts of heat.

The anode and its water jacket act like a distillery. Any impurities in the water will eventually find their way into the water jacket and cause corrosion of the anode. It is essential to use high-purity water with low conductivity, less than 10 mS/cm (millisiemens per centimeter), and to replace the water in the cooling jacket as needed. Efficient heat transfer from the anode surface to the water is a key ingredient for long life. Oil, grease, solder flux residue, and pipe sealant containing silicone compounds must be kept out of the cooling system. While this applies to both vapor-phase and liquid-conduction cooling systems, it is usually more critical in the vapor-phase type. A *sight glass* is usually included in the system to provide a convenient checkpoint for coolant condition.

In regions where freezing temperatures may be experienced, a mixture of distilled water and *ethylene glycol* may be used as the coolant in a water cooled system. The heat transfer of such a mixture is lower than that of pure water, requiring the flow to be increased, typically by 20 to 25 percent. The action of heat and air on ethylene glycol causes the formation of acidic products. Buffers (alkaline salts) may be added with the glycol mixture to neutralize acid forms and prevent corrosion. It should be noted that because they are ionizable chemical salts, buffers cause the conductivity of the coolant to increase.

Experience has shown that the only practical way to ensure good coolant condition is to drain, flush, and recharge the system at least once a year. In practice, a vapor-phase cooling system will remain cleaner longer than a water cooled system because the water is continually being re-distilled. Any contaminants will tend to collect at the bottom of the boiler, where they can be cleaned out during routine maintenance.

7.5.5   Effects of Excessive Dissipation

Proper cooling of the tube envelope and seals is a critical parameter for long tube life. Deteriorating effects that result in shortened life and reduced performance increase with increasing temperature. Excessive dissipation is perhaps the single greatest cause of catastrophic failure in a power tube.

Tubes that operate at VHF and UHF frequencies are inherently subject to greater heating action than devices operated at lower frequencies. This condition is the result of larger RF charging currents into the tube capacitances, dielectric losses, and the tendency of electrons to bombard parts of the tube structure other than the grid and

Figure 7-37 A new, unused 4CX15,000A tube. Contrast the appearance of this device with the tubes that follow. *(Courtesy of Varian Associates.)*

plate at higher frequencies. Greater attention, therefore, must be given to tube cooling at higher frequencies.

Case Histories. Examining a power tube after it has been removed from a transmitter or other type of RF generator can tell a great deal about how well the transmitter-tube combination is working. Contrast the appearance of a new power tube (Figure 7-37) with a component at the end of its useful life. Consider the following examples:

• Figure 7-38. Two 4CX15,000A power tubes with differing anode heat dissipation patterns. Tube (a) experienced excessive heating because of a lack of PA compartment cooling air or excessive dissipation because of poor tuning. Tube (b) shows a normal thermal pattern for a silver-plated 4CX15,000A. Nickel-plated tubes do not show signs of heating because of the high heat resistance of nickel.

Figure 7-38 Anode dissipation patterns on two 4CX15,000A tubes. Tube (a), on the left, shows excessive heating; tube (b), on the right, shows normal wear. *(Courtesy of Econco Broadcast Service.)*

- Figure 7-39. Base heating patterns on two 4CX15,000A tubes. Tube (a) shows evidence of excessive heating because of high filament voltage or lack of cooling air directed toward the base of the device. Tube (b) shows a typical heating pattern with normal filament voltage.
- Figure 7-40. A 4CX5,000A with burning on the screen-to-anode ceramic. Exterior arcing of this type generally indicates a socketing problem, or another condition external to the tube.
- Figure 7-41. The stem portion of a 4CX15,000A that had gone down to air while the filament was on. The grids are burned and melted because of the ionization arcs that subsequently occurred. A failure of this type will trip overload breakers in the transmitter. It is indistinguishable from a shorted tube in operation.

Figure 7-39 Base heating patterns on two 4CX15,000A tubes. Tube (a), on the left, shows excessive heating; tube (b), on the right, shows normal wear. *(Courtesy of Econco Broadcast Service.)*

Figure 7-40 A 4CX5,000A tube that appears to have suffered socketing problems. *(Courtesy of Econco Broadcast Service.)*

Figure 7-41 The interior elements of a 4CX15,000A tube that had gone to air while the filament was lit. *(Courtesy of Econco Broadcast Service.)*

Bibliography

1. *The Care and Feeding of Power Grid Tubes,* Varian Eimac, San Carlos, CA, 1982.
2. Mina and Parry, "Broadcasting with Megawatts of Power: The Modern Era of Efficient Powerful Transmitters in the Middle East," *IEEE Transactions on Broadcasting,* Vol. 35, No. 2, Washington, D.C., June 1989.
3. Jordan, Edward C., *Reference Data for Engineers: Radio, Electronics, Computer and Communications,* Seventh Edition, Howard W. Sams Company, Indianapolis, IN, 1985.
4. *High Power Transmitting Tubes for Broadcasting and Research,* Philips Technical Publication, Eindhoven, The Netherlands, 1988.
5. Gray, T. S., *Applied Electronics,* MIT, 1954.

8

Microwave Power Tubes

8.1 Introduction

Microwave power tubes span a wide range of applications, operating at frequencies from 300 MHz to 300 GHz with output powers from a few hundred watts to more than 10 MW. Applications range from the familiar to the exotic. The following devices are included under the general description of microwave power tubes:

- Klystron, including the *reflex* and *multi-cavity* klystron
- *Multi-stage depressed collector* (MSDC) klystron
- *Klystrode* tube (The Klystrode tube is a registered trademark of Varian Associates.)
- *Traveling-wave tube (TWT)*
- *Crossed-field tube*
- *Coaxial magnetron*
- *Gyrotron*
- Planar triode
- High frequency tetrode

This wide variety of microwave devices has been developed to meet a wide range of applications. Some common uses include:

- UHF-TV transmission
- Shipboard and ground-based radar
- Weapons' guidance systems
- *Electronic counter-measures* (ECM) systems

- Satellite communications
- Tropospheric scatter communications
- Fusion research

As new applications are identified, improved devices are designed to meet the need. Microwave power tube manufacturers continue to push the limits of frequency, operating power and efficiency. Microwave technology is an evolving science.

8.2   Klystron

The klystron is a *linear-beam* device that overcomes the transit-time limitations of a grid-controlled tube by accelerating an electron stream to high velocity before it is modulated. Modulation is accomplished by varying the velocity of the beam, which causes the drifting of electrons into *bunches* to produce RF *space current*. One or more cavities reinforce this action at the operating frequency. The output cavity acts as a transformer to couple the high-impedance beam to a low-impedance transmission line. The frequency response of a klystron is limited by the impedance-bandwidth product of the cavities, which may be extended by stagger tuning or by the use of multiple-resonance filter-type cavities. The klystron is one of the primary means of generating high power at UHF frequencies and above. Output powers for multi-cavity devices range from a few thousand watts to 10 MW or more. The klystron provides high gain and requires little external support circuitry. Mechanically the klystron is relatively simple. It offers long life and requires minimum routine maintenance.

8.2.1   Low Power Klystrons

Although this chapter focuses on the high-power applications of microwave tubes, it is appropriate to discuss briefly the low-power counterparts to multi-element klystrons.

Reflex Klystron. The reflex klystron uses a single cavity resonator to modulate the RF beam and extract energy from it. The construction of a reflex klystron is shown in Figure 8-1. The beam passes through the cavity and is then reflected by a negatively charged electrode, which causes the beam to pass through again in the reverse direction. By varying the applied voltage on the reflector electrode, phasing of the beam can be varied to produce the desired oscillating mode and control the frequency of oscillation. The reflex

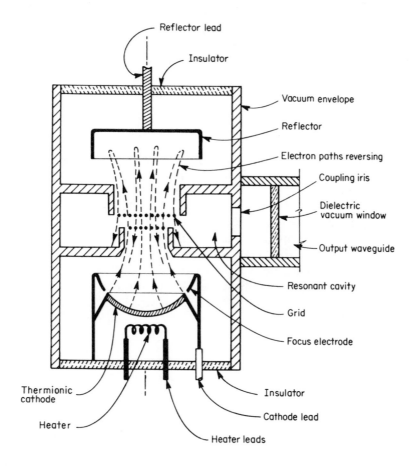

Figure 8-1 Schematic cross section of a reflex klystron oscillator. (*Source: Varian Associates.*)

klystron typically includes a grid to concentrate the electric field so that it may efficiently couple to the electron beam. The reflex klystron is used as a local oscillator, low-power FM transmitter, or test signal source. Reflex tubes are used primarily from 4-40 GHz. Power outputs of 1 W or less are common. The reflex tube is the only klystron in which *beam feedback* is used to produce output energy. In klystrons with more than one cavity, the electron beam passes through each cavity in succession.

Two-Cavity Klystron Oscillator. The two-cavity klystron oscillator is designed for applications requiring moderate power (up to 100 W), stable frequency output and low sideband noise. The device has a coupling iris on the wall between the two cavities. The tube can be frequency-modulated by varying the cathode voltage about the center of the oscillating mode. While more efficient and powerful than the reflex klystron, the two-cavity klystron requires more modulator power. The two-cavity klystron is typically used in Doppler radar systems.

Two-Cavity Klystron Amplifier. Similar in design to the two-cavity oscillator, the two-cavity klystron amplifier provides limited power output (10 W or less) and moderate gain (about 10 dB). A driving signal is coupled into the input cavity, which produces velocity modulation of the beam. After the drift space, the density-modulated beam induces current in the output resonator. Electrostatic focusing of the beam is common. The two-cavity klystron finds only limited applications because of its restrictions on output power and gain. For many applications, solid-state amplifiers are a better choice.

8.2.2   Multi-Cavity Klystron

The basic mechanical construction of a multi-cavity klystron is shown in Figure 8-2. A filament, heated by an electric current, in turn heats a cathode coated to emit electrons when it reaches a sufficiently high temperature. Negatively charged electrons attracted by a positively charged anode pass through the first cavity of the klystron. Microwaves in the cavity interact with the electrons to vary their velocity. The electrons then pass through a narrow passage called a *drift tube*. In the drift tube, the electrons tend to bunch up, as the faster ones catch up with the slower ones. At the place in the drift tube where the bunching is most pronounced, the electrons enter a second cavity, where stronger microwaves are excited and amplified in the process.

8.2.3   Theory of Operation

The principle of *velocity modulation* is the basis on which the klystron operates. This concept describes how a continuous electron beam is converted into a density-modulated beam at microwave frequencies. Velocity modulation of a long continuous electron beam gives the klystron an important feature — high gain. Klystron design also uses a collector that can be made large and independent of the

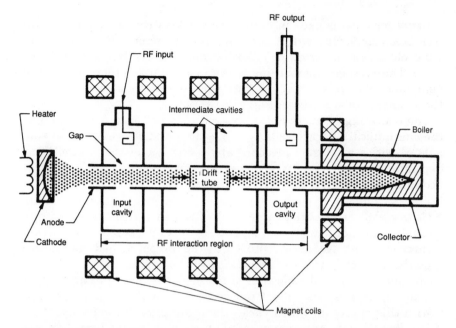

Figure 8-2 Basic mechanical structure of a multi-cavity klystron. The "boiler" shown in the diagram is part of the heat removal system. (*Source: Broadcast Engineering Magazine.*)

operating frequency. In addition, the power density is kept low throughout the tube, which provides long life and high reliability.

The RF input signal is coupled into the input cavity by an inductive loop. The RF voltage impressed across the cavity's capacitive gap applies an RF component to the beam current. The intermediate cavities enhance this process until, at the output cavity gap, the electron beam is formed into more or less tight bunches. By judicious design of the output cavity, output cavity gap and output loading loop, the energy in the bunched beam can be efficiently extracted and delivered to the load.

Each cavity tuned to the operating frequency adds about 20 dB gain to the 10 dB gain offered by the two-cavity klystron amplifier. Overall gains of 60 dB are practical. Cavities may be tuned to either side of resonance to broaden the operating bandwidth of the device. Klystrons with up to eight cavities have been produced. Operating power for CW klystrons range up to 1 MW per device, and as much as 50 MW per device for pulsed applications.

The primary physical advantage of the klystron over a grid-based power tube is that the cathode-to-collector structure is virtually independent of transit-time effects. Therefore, the cathode can be made large and the electron beam density kept low.

Beam focusing is used in multi-element klystrons to keep the electron beam uniformly small. Focusing may be accomplished by one or more electrostatic *lenses,* or external magnetic fields placed parallel to the beam. Magnetic focusing is the most common method. Permanent magnets are used at operating powers of about 5 kW or less. Electromagnets are used at higher powers.

The operating frequency of a klystron may be fixed, determined by the mechanical characteristics of the tube and its cavities, or tunable. Cavities are tuned mechanically using one of several methods, depending on the operating power and frequency. Tuning may be accomplished by the following:

- Cavity wall deformation. One wall of the cavity consists of a thin diaphragm that is moved in and out by a tuning mechanism. About three percent frequency shift may be accomplished using this method, which varies the inductance of the cavity.
- Movable cavity wall. One wall of the cavity is moved in and out by a tuning mechanism. About ten percent frequency shift is possible with this approach, which varies the inductance of the cavity.
- Paddle element. An element inside the cavity moves perpendicular to the beam and adds capacity across the interaction gap. A tuning range of about 25 percent is provided by this approach.
- Combined inductive-capacitive tuning. Incorporates a combination of the previous methods. Tuning variations of 35 percent are possible.

Each of these tuning methods may be used whether the cavity is inside or outside the vacuum envelope of the tube. Generally speaking, however, tubes that use external cavities provide more adjustment range, usually on the order of 35 percent. Bandwidth may be increased by stagger tuning the cavities, at the expense of gain.

High-conversion efficiency requires the formation of electron bunches, which occupy a small region in velocity space, and the formation of *inter-bunch regions* with low electron density. The latter is particularly important because these electrons are phased to be accelerated into the collector at the expense of the RF field. Studies show that energy loss as a result of an electron accelerated into the collector can exceed the energy delivered to the field by an equal but properly phased electron. Herein lies a key in improving the efficiency of the klystron — recover a portion of this wasted energy.

Klystrons are cooled by air or liquid for powers up to 5 kW. Tubes operating in excess of 5 kW are usually water- or vapor-cooled.

Perveance. The *perveance* of an electron beam is defined by the equation:

$$I_b = k\ (E_b^{3/2})$$

Where:
I_b = beam current in A
E_b = beam voltage in kV
k = *microperveance* (a constant)

Microperveance is a function of the dimensions of the electron gun. It has been recognized for some time that klystrons with lower perveance are more efficient than tubes that operate at low beam voltage and high current. High efficiency in low-perveance beams is mainly due to the charge density in a bunched electron beam, which is limited by the mutual repulsion of the electrons in that bunch. This mutual repulsion tends to deform the bunches, resulting in low efficiency.

Phased Electrons. Experience has shown that by properly phasing the second-harmonic fields of a klystron, a favorable electron density distribution pattern can be established at the output gap. The result is the generation of additional RF energy.

A phase-space diagram for a high-power klystron is shown in Figure 8-3. The curves represent a plot of the relative phase of the reference electrons as a function of axial distance along the tube. Electrons having negative slope have been decelerated. Electrons having

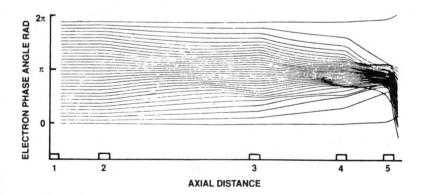

Figure 8-3 Plot of the relative phase of the reference electrons as a function of axial distance in a high-efficiency klystron. (*Source: Broadcast Engineering Magazine.*)

positive slope have been accelerated with respect to a non-accelerated electron parallel to the axis. The diagram shows how the electrons are nicely grouped at the output cavity gap, while the inter-bunch regions are relatively free of electrons. This interaction can be viewed another way, as shown in Figure 8-4, which plots the normalized RF beam currents as a function of distance along the tube. The curves show that the fundamental component of the plasma wave has a negative slope at the third gap. This normally would be a poor condition, but because of the drift of inter-bunch electrons, the fundamental current peaks at nearly 1.8 times the dc beam current. The theoretical limit for perfect bunching in a delta function is 2. The second-harmonic of the plasma wave also peaks at the output gap, which adds to the conversion efficiency.

Types of Devices. Klystrons can be classified by the following basic parameters:

* Power-operating level. Klystrons are available ranging from a few hundred watts to more than 10 MW.
* Operating frequency. Klystrons are typically used over the frequency range of 300 MHz to 40 GHz.

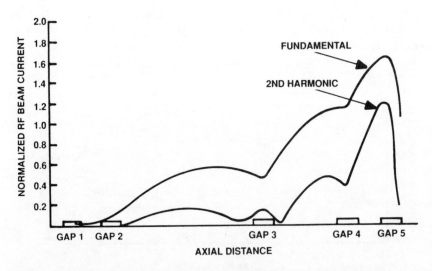

Figure 8-4 Plot of the normalized RF beam currents as a function of distance along the length of a high-efficiency klystron. (*Source: Broadcast Engineering Magazine.*)

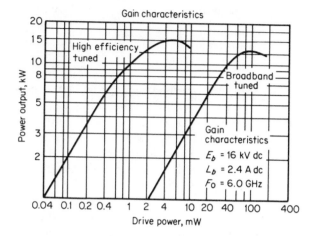

Figure 8-5 Typical gain, output power, and drive requirements for a klystron. (*Source:* Fink and Christiansen, *Electronics Engineers' Handbook,* McGraw-Hill, New York, 1989.)

- Number of cavities. The number of resonant cavities may range from one to five or more. Further, the cavities may be *integral* or *external* to the vacuum envelope of the device.

The klystron is a true linear amplifier from zero signal level to up to 2-3 dB below saturated output. Figure 8-5 shows a typical transfer characteristic for a klystron. RF modulation is applied to the input drive signal. Amplitude modulation is typically limited to the linear portion of the gain transfer characteristic (class A operation). The result is low efficiency, because the beam power is always on. For applications requiring frequency modulation, the drive power is set for saturated output.

Pulse modulation of the klystron is obtained by applying a negative rectangular voltage to the cathode, instead of a dc voltage. The RF drive, set to a saturation value, is usually pulsed on for a slightly shorter time than the beam pulse.

Because of the high-power levels typically used at UHF frequencies, device efficiency is a critical parameter. Klystrons are usually rated in terms of *saturated efficiency,* determined by dividing the saturated RF output power by the dc input power. Saturated efficiency governs the maximum *peak-of-sync efficiency* available when *beam pulsing* techniques are employed for UHF television service. Peak-of-sync efficiency is the commonly used *figure of merit* (FOM) expression, defined as the peak-of-sync output power divided by the dc input power.

8.2.4   Beam Pulsing

Beam pulsing is a common method of improving the efficiency a of broadband linear klystron amplifier. Depending on the transmitted waveform, efficiency may be boosted by 25 percent or more. This technique is typically used in UHF-TV transmitters to reduce visual klystron beam dissipation during video portions of the transmission. *Sync pulsing*, as the technique is commonly referred, is accomplished by changing the operating point of the tube during the synchronizing interval, when peak power is required, and returning it to a linear transfer characteristic during the video portion of the transmission. This control is accomplished through the application of a voltage to an electrode placed near the cathode of the klystron. Biasing toward cathode potential increases the beam current, while biasing toward ground (collector potential) decreases beam current.

In the composite television waveform, video information occupies 75 percent of the amplitude and sync occupies the remaining 25 percent. The *tip of sync* represents the peak power of the transmitted waveform. Black (the *blanking level*) represents 56 percent of the peak power. If the blanking level could be made to represent 100 percent modulation and the sync pulsed in, as in a radar system, efficiency would be increased significantly. Unfortunately, the color burst signal extends 50 percent into the sync region, and any attempt to completely pulse the sync component would distort the color-burst reference waveform. Sync pulsing is, therefore, limited to 12.5 percent above black to protect color-burst.

Two common implementations of beam pulsing can be found: *modulating anode* (mod-anode) pulsing and *annular beam control electrode* (ACE) pulsing. The mod-anode system utilizes an additonal electrode after the cathode to control beam power. The ACE-type tube operates on a similar principal, however, the annular ring is placed such that the ring encircles the electron beam. Because of the physical design, pulsing with the ACE-type gun is accomplished at a much lower voltage than with a mod-anode device. The ACE element, in effect, grid-modulates the beam.

In theory, the amount of beam-current reduction achievable, and the resulting efficiency improvement, are independent of whether mod-anode or ACE-type pulsing are used. In practice, however, differences are noted. With existing mod-anode pulsers, an efficiency improvement of about 19 percent in beam current over non-pulsed operation may be achieved. A beam reduction of 30 to 35 percent may be achieved through use of an ACE-equipped tube. The ACE electrode, positioned in the gun assembly, is driven by a negative-going narrow-band video signal of a few hundred volts peak. The annular ring varies the beam density through a *pinching action* that effectively reduces the cross-sectional area of the stream of electrons emitted by the cathode. The klystron thus operates in a quasi-class A-B condition rather than the normal class A (for linear television service).

A peak-of-sync FOM for an integral cavity UHF-TV klystron without ACE or equivalent control is 0.67-0.68. By using an ACE-type tube, the FOM may be increased to 0.80. Similar improvements in efficiency can be realized for external-cavity klystrons. A typical switching-type mod-anode pulser is shown in Figure 8-6.

Pulsing is not without its drawbacks, however. The greater the pulsing, the greater the pre-correction required from the modulator. Pre-correction is needed to compensate for non-linearities of the klystron transfer characteristic during the video modulation period. Level-dependent RF phase pre-correction may also be required. Switching between different klystron characteristics produces phase modulation of the visual carrier. If not corrected, this phase modulation may result in intercarrier "buzz" in the received audio. These and other considerations limit the degree of pulsing that may be achieved on a reliable basis.

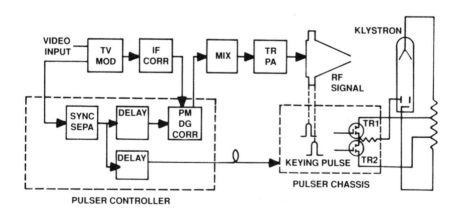

Figure 8-6 Block diagram of a switching mod-anode pulser. (*Source: Broadcast Engineering Magazine.*)

8.2.5　Integral Versus External Cavity

In an integral-cavity klystron, the resonant cavities are located within the vacuum envelope of the tube. In an external-cavity klystron, the cavities are located outside the envelope in a mechanical assembly that wraps around the drift tube.

Fundamentally, klystron theory applies equally to integral- and external-cavity tubes. In both cases, a velocity-modulated electron beam interacts with multiple-resonant cavities to provide an amplified output signal. The resonant cavity *interaction gap* and drift length requirements for optimum performance, including conversion efficiency, are independent of whether the tuning mechanism is inside or outside the vacuum envelope. High-efficiency integral- and external-cavity klystrons have been designed to provide maximum conversion efficiency consistent with signal bandwidth requirements. The saturated conversion efficiency is essentially identical for integral- or external-cavity klystrons *with equal numbers of resonant cavities.*

Number of Cavities. A debate has been underway for a number of years in the UHF television industry over the relative benefits of integral-versus-external cavity klystrons. Discussion centers on operating efficiency, life expectancy, and replacement device cost. When comparing the efficiency of integral-vs.-external cavity klystrons for UHF-TV applications, however, the question really boils down to the number of cavities (four or five) used in the device. Four-cavity external tubes are standard. It is usually not practical to produce high-power five-cavity external tubes for mechanical and electrical (voltage standoff) reasons.

Integral devices permit the addition of a fifth cavity to the design for two primary reasons. First, the device itself is mechanically more robust. The addition of a fifth cavity to an external klystron adds length and weight to the device. Because of the size of the drift tube in an external device designed for UHF-TV frequencies, the additional weight is hard to support. Second, because the cavities are enclosed in a vacuum in the integral design, voltage standoff problems are greatly reduced.

The tuning mechanism of an integral-cavity klystron is enclosed in a rigid steel shell. In the external-cavity unit the tuning mechanism is in air. Within the region of the tuning mechanism, RF fields can reach high levels, especially at the high end of the UHF-TV band. The highest energy field occurs in the area of the last cavity. Because of the high dielectric properties of a vacuum, dielectric breakdown is less of a problem in the integral design, which places the tuning mechanisms within a vacuum envelope. This situation is of particular concern at the high end of the UHF band, where the spacing of tuning elements is closer.

Four-cavity integral-type klystrons are manufactured, but for power outputs of 30 kW or less. 60-kW-integral klystrons, the bread and butter of UHF-TV broadcasting, are produced with five-cavity designs.

The physics of the integral- and external-cavity klystrons are essentially the same. When comparing integral and external units of like design, that is with the same number of cavities, performance should be identical. The two units follow the same laws of physics, and use basically the same components up to the beam stick. It is when manufacturers take advantage of the relative merits of each design that differences in performance are realized.

Efficiency. For the sake of comparison, the data presented in this section will assume: (1) a power level of 60 kW, (2) all integral- cavity devices utilize five cavities, and (3) all external-cavity devices utilize four cavities.

The five-cavity integral *S-tuned* klystron is inherently at least 20 percent more efficient than a four-cavity tube. (*S-tuning* refers to the method of stagger-tuning the cavities.) The five-cavity klystron is generally specified by the manufacturer for a minimum efficiency of 52 percent (saturated efficiency). Typical efficiency is 55 percent. Four-cavity devices are characteristically specified at 42 percent minimum, and 45 percent typical.

This efficiency advantage is possible because the fifth cavity of the integral design permits tuning patterns that allow maximum transfer of RF energy while maintaining adequate bandpass response. Tighter bunching of electrons in the beamstick, a function of the number of cavities, also contributes to the higher efficiency operation. In actuality, the fifth cavity allows design engineers to trade gain for efficiency. Still, the five-cavity tube has significantly more gain than a four-cavity device. The five-cavity unit, therefore, requires less drive, which simplifies the driving circuit. A five-cavity klystron requires less than 25 W of drive power, while a four-cavity tube needs as much as 90 W.

Under pulsed operation, approximately the same reduction in beam current is realized with both integral- and external-cavity klystrons. When comparing peak-of-sync FOM, the efficiency differences will track. There is, fundamentally, no reason that one type of klystron should pulse differently than the other.

Performance Tradeoffs. It is a designer's choice whether to build a transmitter with the klystron cavities located inside or outside the vacuum envelope. There are benefits and drawbacks to each approach.

When the cavity resonators are a part of the tube, the device becomes more complicated and more expensive. But, the power-generating system is all together in one package, and that simplifies installation significantly.

When the resonator is separate from the tube (an external-cavity device), it can be made with more *compliance* (greater room for adjustment). Consequently, a single device may be used over a wider range of operating frequencies. In terms of UHF-TV, a single external cavity device may be tuned for operation over the entire UHF-TV band. This feature is not possible if the resonant cavities are built into the device. To cover the entire UHF-TV band, three integral cavity tubes are required. The operational divisions are:

- Channels 14-29 (470-566 MHz)
- Channels 30-51 (566-698 MHz)
- Channels 52-69 (698-746 MHz)

This practical limitation to integral-cavity klystron construction may be a drawback for some facilities. For example, it is not uncommon for group operations to share one or more spare klystrons. If the facilities have operating frequencies outside the limits of a single integral device, it may be necessary to purchase more than one spare. Also, when the cavities are external, the resonators are in air and can be accessed to permit fine adjustments of the tuning stages for peak efficiency.

The advantages of tube changing when an integral device is used are significant. Typical tube-change time for an integral klystron is one hour, as opposed to four to six hours for an external device. The level of experience of the technician is also more critical when an external-type device is being changed. Tuning procedures must be carefully followed by maintenance personnel to avoid premature device failure.

Reliability. Klystron lifespan data has been developed by the manufacturers of both integral- and external-cavity tubes. Usually these figures are based on surveys of users and represent actual installed life. It is difficult to separate tube failures resulting from external forces, such as lightning or power line transients, and true tube faults.

Although the data from tube manufacturers varies, one representative lifespan survey shows an average tube life of 41,260 hours for an integral-cavity 60-kW device. Most published lifetime data for external-cavity designs of the same power level indicate shorter life expectancy (about 30,000 hours). This comparison can somewhat be misleading, however, because more integral-cavity 60-kW tubes (designed principally for UHF-TV) have been installed in sockets in the U.S. — the primary basis of comparison — for a longer period of time than external-cavity tubes. Therefore, more life-cycle data is available for integral-cavity devices (as of this writing). An argument can reasonably be made that after enough statistical data has been gathered on both types of devices, the life expectancy of each tube will be shown to be essentially the same.

It is fair to acknowledge that the output window tends to be more of a problem with the external-cavity tube when load mismatches are present. The window experiences greater stress because it is in close proximity to the output of the device. It is, in fact, within the cavity resonator.

It should be emphasized that because of the way life-cycle data is gathered, the projected number of operating hours is not directly a function of the tube itself, but rather of the tube and transmitter combination operating in the on-site environment. From time to time, manufacturing problems in a product batch will result in a given device being more susceptible to failure caused by overload. These occurrences, however, are anomalies that cannot reasonably be considered in an analysis of inherent tube life.

8.3   Traveling-Wave Tube[1]

The traveling-wave tube (TWT) is a linear-beam device finding extensive applications in communications and research. Power levels range from a few watts to 10 MW. Gain ranges from 40-70 dB for small drive signals. (Figure 8-7 shows the basic elements of a TWT.) The TWT consists of four basic elements:

- *Electron gun.* The gun forms a high-current-density beam of electrons that interact with a wave traveling along the RF circuit to

---

[1] Portions of Section 8.3 (Traveling-Wave Tubes) and Section 8.4 (Crossed-Field Tubes) were adapted from *The Electronics Engineers Handbook, Third Edition,* Section 9, "UHF and Microwave Devices," Donald G. Fink and Donald Christiansen, Editors.

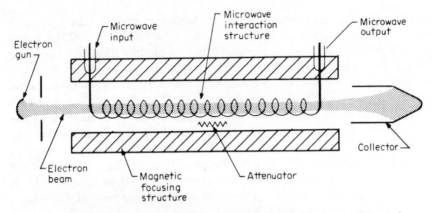

Figure 8-7 Basic elements of a traveling-wave tube. (*Source:* Fink and Christiansen, *Electronics Engineers' Handbook,* McGraw-Hill, New York, 1989.)

increase the amplitude of the RF signal. In a typical application, electrons are emitted from a cathode and converged to the proper beam size by focusing electrodes.

- *RF interaction circuit.* The RF wave is increased in amplitude as a result of interaction with the electron beam from the gun. The fundamental principal on which the TWT operates is that an electron beam moving at approximately the same velocity as an RF wave traveling along a circuit gives up energy to the RF wave.
- *Magnetic electron-beam focusing system.* The beam size is maintained at the proper dimensions through the interaction structure by the focusing system. This may be accomplished by using either a permanent magnet or an electromagnetic focusing element.
- *Collector.* The electron beam is received at the collector after it has passed through the interaction structure. The remaining beam energy is dissipated in the collector.

The primary differences between types of TWT devices involves the RF interaction structure employed. In the previous figure, the interaction structure was a helix. A variety of other structures may be employed, depending on the operating power and frequency.

Three common approaches are used to provide the needed magnetic-beam focusing. Illustrated in Figure 8-8, they are:

- Electromagnetic focusing. Used primarily on high-power tubes where tight beam focusing is required (Figure 8-8(a)).
- Permanent magnet focusing. Used where the interaction structure is short (Figure 8-8(b)).

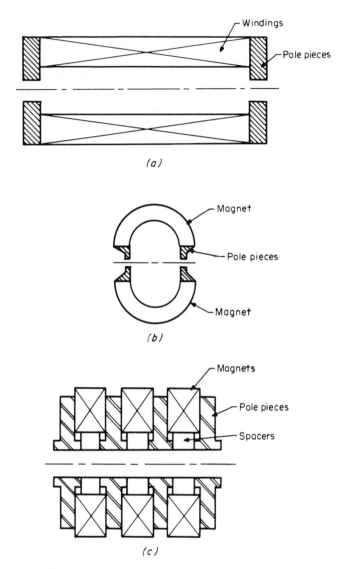

Figure 8-8 Magnetic focusing for a TWT: (a) solenoid-type; (b) permanent-magnet-type; (c) periodic permanent-magnetic structure. (*Source:* Fink and Christiansen, *Electronics Engineers' Handbook*, McGraw-Hill, New York, 1989.)

- Periodic permanent magnet focusing. Used on most helix TWT and coupled-cavity tubes. The magnets are arranged with alternate polarity in successive cells along the interaction region (Figure 8-8(c)).

8.3.1   Theory of Operation

The interaction structure acts to slow the RF signal so that it travels at approximately the same speed as the electron beam. Electrons enter the structure during both positive and negative portions of the RF cycle. Electrons entering during the positive portion are accelerated; those entering during the negative portion are decelerated. The result is the creation of *electron bunches* that produce an alternating current superimposed on the dc beam current. This alternating current induces the growth of an RF *circuit wave* that encourages even tighter electron bunching. One or more *severs* are included to absorb reflected power that travels in a backward direction on the interaction circuit. This reflected power is the result of a mismatch between the output port and the load. Without the sever, regenerative oscillations could occur. At a given frequency, a particular level of drive power will result in maximum bunching and power output. This operating point is referred to as *saturation.*

Interaction Circuit. The key to TWT operation lies in the interaction element. Because RF waves travel at the speed of light, a method must be provided to slow down the forward progress of the wave to approximately the same velocity as the electron beam from the cathode. The beam speed of a TWT is typically 10 to 50 percent the speed of light, corresponding to cathode voltages of 4 to 120 kV. Two mechanical structures are commonly used to provide the needed slowing of the RF wave:

- *Helix circuit.* The helix is used where bandwidths of an octave or more are required. Over this range the velocity of the signal carried by the helix is basically constant with frequency. Typical operating frequencies range from 500 MHz to 40 GHz. Operating power, however, is limited to a few hundred watts. TWTs intended for higher frequency operation may use a variation of the helix, shown in Figure 8-9. The *ring-loop* and *ring-bar* designs permit peak powers of hundreds of kilowatts. The average power, however, is about the same as a conventional helix because the structure used to support the interaction circuit is the same.
- *Coupled-cavity circuit.* The coupled-cavity interaction structure permits operation at high-peak and average power levels, and moderate bandwidth (10 percent is typical). TWTs using coupled-cavity structures are available at operating frequencies from 2-100 GHz. The basic design of a couple-cavity interaction circuit is shown in Figure 8-10. Resonant cavities, coupled through slots cut in the

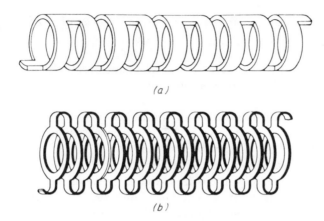

(a)

(b)

Figure 8-9 Helix structures for a TWT: (a) ring-loop circuit; (b) ring-bar circuit. (*Source:* Fink and Christiansen, *Electronics Engineers' Handbook,* McGraw-Hill, New York, 1989.)

cavity end walls, resemble a folded waveguide. Two basic schemes are used: the *cloverleaf* and the *single-slot space harmonic* circuit.

The cloverleaf, also known as the *forward fundamental* circuit, is used primarily on high-power tubes. The cloverleaf provides operation at up to 3 MW peak power and 5 kW average at S-band frequencies.

The single-slot space harmonic interaction circuit is more common than the cloverleaf. The mechanical design is simple, as shown in the figure. The single-slot space harmonic structure typically provides

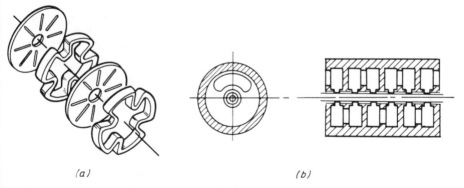

(a)   *(b)*

Figure 8-10 Coupled-cavity interaction structures: (a) forward fundamental circuit or "cloverleaf"; (b) single-slot space harmonic circuit. (*Source:* Fink and Christiansen, *Electronics Engineers' Handbook,* McGraw-Hill, New York, 1989.)

peak power of up to 50 kW and average power of 5 kW at X-band frequencies.

Pulse Modulation. The electron beam from the gun may be pulse-modulated using one of four methods:

- *Cathode pulsing.* The cathode is pulsed in a negative direction with respect to the grounded anode. This approach requires that the full beam voltage and current be switched.
- *Anode pulsing.* This approach is similar to cathode pulsing, except that the full beam voltage is switched between cathode potential and ground. The *current* switched, however, is only that value intercepted on the anode. Typically the intercepted current is a few percent of the full beam potential.
- *Focus electrode pulsing.* If the focus electrode, which normally operates at or near cathode potential, is biased negatively with respect to the cathode, the beam will be turned off. The voltage swing required is typically 1/3 of the full cathode voltage. This approach is attractive because the focus electrode draws essentially no current, making implementation of a switching modulator relatively easy.
- *Grid pulsing.* The addition of a grid to the cathode region permits control of beam intensity. The voltage swing required for the grid, placed directly in front of the cathode, is typically 5 percent of the full beam potential.

8.3.2 Operating Efficiency

Efficiency is not one of the TWT's strong points. Early traveling wave tubes offered only about 10 percent dc-to-RF efficiency. Wide bandwidth and power output are where the TWT shines. TWT efficiency may be increased in two basic ways: (1) *collector depression* for a single-stage collector, or (2) use of a *multi-stage collector.*

Collector depression refers to the practice of operating the collector at a voltage lower than the full beam voltage. This introduces a potential difference between the interaction structure and the collector, through which electrons pass. The amount by which a single-stage collector can be depressed is limited by the remaining energy of the slowest electrons. In other words, the potential reduction can be no greater than the amount of energy of the slowest electrons, or they will be turned around and reenter the interaction structure causing oscillation.

By introducing multiple depressed collector stages, still greater efficiency can be realized. This method provides for the collection of the slowest electrons at one collector potential, while allowing those with more energy to be collected on other stages that are depressed still further (see Figure 8-11).

Cooling. Cooling of a low power TWT is accomplished by clamping the tube to a metal baseplate, mounted in turn, on an air- or liquid-cooled heat sink. Coupled-cavity tubes below 1 kW average power are convection-cooled by circulating air over the entire length of the device. Higher power coupled-cavity tubes are cooled by circulating liquid over the tube body and collector.

8.3.3 Operational Considerations

While TWTs offer numerous benefits to the end-user, they are not without their problems.

Intermodulation Distortion. TWTs are susceptible to intermodulation distortion when multiple carriers are introduced, as in double-illuminated or multiple-SCPC (*single channel per carrier*) sat-

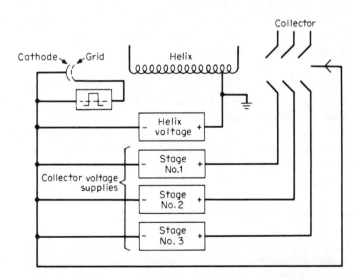

Figure 8-11 Power supply configuration for a multi-stage depressed collector TWT. (*Source:* Fink and Christiansen, *Electronics Engineers' Handbook,* McGraw-Hill, New York, 1989.)

ellite transponders. These intermodulation (IM) products may be found at frequencies that are displaced from the fundamental carriers by the difference in the frequency between them. When multiple carriers are present, the potential for IM exists. This potential is reduced by operating the TWT below saturation. Power must be reduced (backed off) in proportion to the number of carriers and their relative power.

Second-Harmonic Content. Because of the wide bandwidth and non-linear operating characteristics under saturation conditions, a TWT can generate significant second-harmonic energy. It is not uncommon to measure second-harmonic energy at the output of a TWT that is down only 10 dB from the operating carrier. Reduction of harmonic content usually involves injecting a coherent harmonic signal with controlled phase and amplitude along with the fundamental carrier so that they interact, with the end result that harmonics are minimized.

AM/PM Conversion. AM/PM conversion is the change in phase angle between the input and output signals as the input varies in amplitude. The root cause of this distortion in a TWT centers on the reduction of electron beam velocity as the input signal level increases. This causes a greater energy exchange between the electron beam and the RF wave. At a level 20 dB below the input power required for saturation, AM/PM conversion is negligible. At higher levels, AM/PM distortion may increase.

Phase Variation. When the velocity of the electron beam in the TWT is changed, the phase of the output signal will also vary. The primary causes of beam velocity variations include changes in one or more of the following: (1) cathode temperature, (2) grid voltage, (3) anode voltage, and (4) cathode voltage. The TWT power supply must be well-regulated (to less than 0.2 percent) to prevent beam velocity changes that may result in output signal phase variations.

8.4   Crossed-Field Tubes

A crossed-field microwave tube is a device that converts dc into microwave energy using an electronic energy-conversion process. These devices differ from *beam tubes* in that they are *potential-energy converters,* rather than kinetic-energy converters. The term *crossed-field* is derived from the orthogonal characteristics of the dc electric field supplied by the power source and the magnetic field required for beam focusing in the interaction region. This magnetic field is typically supplied by a permanent-magnet structure. Such devices are also referred to *M-tubes*.

Practical devices based on the crossed-field principles fall into two broad categories:

- *Injected-beam crossed-field tubes.* The electron stream is produced by an electron gun located external to the interaction region, similar to a TWT. The concept is illustrated in Figure 8-12.
- *Emitting-sole tubes.* The electron current for interaction is produced directly within the interaction region by secondary electron emissions, which result when some electrons are driven to the negative electrode and allowed to strike it. The negative electrode is formed using a material capable of producing significant secondary-emission electrons. The concept is illustrated in Figure 8-13.

8.4.1 Magnetron

The magnetron encompasses a class of devices finding a wide variety of applications. Pulsed magnetrons have been developed that cover frequency ranges from the low UHF band to 100 GHz. Peak power from a few kilowatts to several megawatts has been obtained. Typical overall efficiencies of 30 to 40 percent may be realized, depending on the power level and operating frequency. CW magnetrons have also been developed, with power levels of a few hundred watts in a tunable tube, and up to 25 kW in a fixed-frequency device. Efficiencies range from 30 percent to as much as 70 percent.

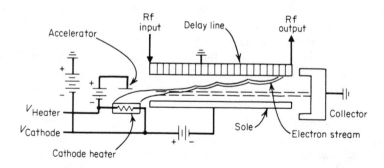

Figure 8-12 Linear injected beam microwave tube. (*Source:* Fink and Christiansen, *Electronics Engineers' Handbook,* McGraw-Hill, New York, 1989.)

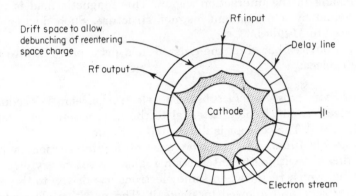

Figure 8-13 Reentrant emitting-sole crossed-field amplifier tube. (*Source:* Fink and Christiansen, *Electronics Engineers' Handbook,* McGraw-Hill, New York, 1989.)

The magnetron operates electrically as a simple diode. Pulsed modulation is obtained by applying a negative rectangular voltage waveform to the cathode with the anode at ground potential. Operating voltages are less critical than for beam tubes; line-type modulators are often used to supply pulsed electric power. The physical structure of a conventional magnetron is shown in Figure 8-14.

High-power pulsed magnetrons are used primarily in radar systems. Low-power pulsed devices find applications as beacons. Tunable CW magnetrons are used in ECM (electronic countermeasures) applications. Fixed-frequency devices are used as microwave heating sources.

Tuning of conventional magnetrons is accomplished by moving capacitive tuners, or by inserting symmetrical arrays of plungers into the inductive portions of the device. Tuner motion is produced by a mechanical connection through flexible bellows in the vacuum wall. Tuning ranges of 10 to 12 percent of bandwidth are possible for pulsed tubes, and as much as 20 percent for CW tubes.

Coaxial Magnetron. The frequency stability of a conventional magnetron is affected by variations in the load impedance and by cathode current fluctuations. Depending on the extent of these two influences, the magnetron may occasionally fail to produce a pulse. The coaxial magnetron minimizes these effects by using the anode geometry shown in Figure 8-15. Alternate cavities are slotted to provide coupling to a surrounding coaxial cavity.

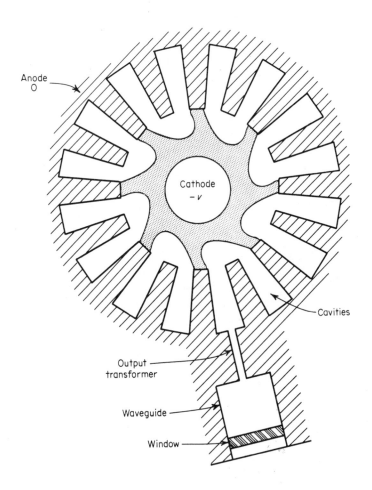

Figure 8-14 Conventional magnetron structure. (Source: Fink and Christiansen, *Electronics Engineers' Handbook,* McGraw-Hill, New York, 1989.)

The oscillating frequency is controlled by the combined-vane system and the resonant cavity. Tuning may be accomplished through the addition of a movable end plate in the cavity, as shown in Figure 8-16.

Frequency-Agile Magnetron. Tubes have been developed for specialized radar and ECM applications that permit rapid tuning of

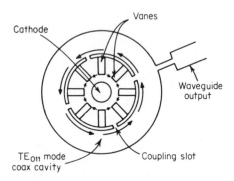

Figure 8-15 Structure of a coaxial magnetron. (*Source:* Fink and Christiansen, *Electronics Engineers' Handbook,* McGraw-Hill, New York, 1989.)

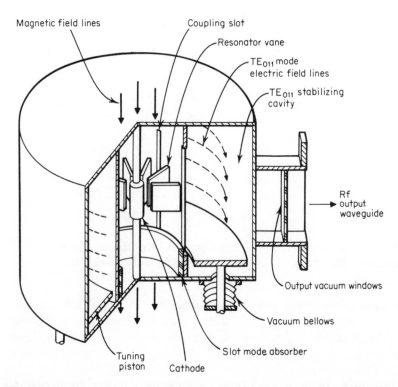

Figure 8-16 Structure of a tunable coaxial magnetron. (*Source:* Fink and Christiansen, *Electronics Engineers' Handbook,* McGraw-Hill, New York, 1989.)

the magnetron. A conventional device may be tuned using one of the following methods:

- A rapidly rotating capacitive element. Tubes of this type are referred to as *spin-tuned magnetrons.*
- A hydraulic-driven tuning mechanism. Tubes of this type are referred to as *mechanically tuned magnetrons.*

Electronic tuning of magnetrons has been demonstrated in the laboratory, with tuning rates as large as several megahertz per microsecond.

Amplitron. The Amplitron is essentially a magnetron with two external couplings that enable amplifier operation. The device is characterized by high power, broad bandwidth, high efficiency, and low gain. The Amplitron acts as a low-loss passive transmission line in the absence of high voltage. The conversion efficiency of an Amplitron can be as high as 85 percent. Pulsed peak-power outputs of more than 1 MW may be realized.

8.4.2 Gyrotron

The *Gyrotron* is a *cyclotron resonance maser.* (Maser is an acronym for *microwave amplification by simulated emission of radiation.* Maser is a general class of microwave amplifier based on molecular interaction with electromagnetic radiation.) The device includes a cathode, collector and circular waveguide of gradually varying diameter. Electrons are emitted at the cathode with small variations in speed. The electrons are then accelerated by an electric field and guided by a static magnetic field through the device. The nonuniform induction field causes the rotational speed of the electrons to increase. The linear velocity of the electrons, as a result, decreases. The interaction of the microwave field within the waveguide and the rotating (helical) electrons causes bunching similar to the bunching within a klystron. A decompression zone at the end of the device permits decompression and collection of the electrons.

The power available from a Gyrotron is 100 times greater than that possible from a classical microwave tube at the same frequency. Power output of 12 kW is possible at 100 GHz, with 30 percent efficiency. At 300 GHz, up to 1.5 kW may be realized, but with only 6 percent efficiency.

8.5 Grid Vacuum Tubes

The physical construction of a vacuum tube causes the output power and available gain to decrease with increasing frequency. The principal limitations faced by grid-based devices include the following:

- Physical size. Ideally, the RF voltages between electrodes should be uniform, however, this condition cannot be realized unless the major electrode dimensions are significantly less than 1/4-wavelength at the operating frequency. This restriction presents no problems at VHF frequencies, however, as the operating frequency increases into the microwave range, severe restrictions are placed on the physical size of individual tube elements.
- Electron transit-time. Interelectrode spacing, principally between the grid and the cathode, must be scaled inversely with frequency to avoid problems associated with electron transit-time. Possible adverse conditions include: (1) excessive loading of the drive source, (2) reduction in power gain, (3) back-heating of the cathode as a result of electron bombardment, and (4) reduced conversion efficiency.
- Voltage standoff. High power tubes operate at high voltages. This presents significant problems for microwave vacuum tubes. For example, at 1 GHz the grid-cathode spacing must not exceed a few mils. This places restrictions on the operating voltages that may be applied to individual elements.
- Circulating currents. Substantial RF currents may develop as a result of the inherent interelectrode capacitances and stray inductances/capacitances of the device. Significant heating of the grid, connecting leads, and vacuum seals may result.
- Heat dissipation. Because the elements of a microwave grid tube must be kept small, power dissipation is limited.

Still, some grid vacuum tubes find applications at high frequencies. Planar triodes are available that operate at several gigahertz, with output powers of 1 to 2 kW in pulsed service. Efficiency (again for pulsed applications) ranges from 30 to 60 percent, depending on the frequency.

8.5.1 Planar Triode

A cross-sectional diagram of a planar triode (7289) is shown in Figure 8-17. The envelope is made of ceramic, with metal members penetrating the ceramic to provide for connection points. The metal

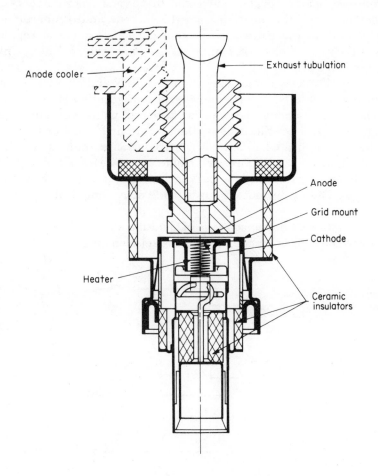

Figure 8-17 Cross-section of a 7289 planar triode. *(Source: Varian Associates.)*

members are shaped either as disks or as disks with cylindrical projections.

The cathode is typically oxide-coated and indirectly-heated. The key design objective for a cathode is high emission density and long tube life. Low-temperature emitters are preferred because high cathode temperatures typically result in more evaporation and shorter life.

The grid of the planar triode represents perhaps the greatest design challenge for tube manufacturers. Close spacing of small-sized elements is needed, at tight tolerances. Good thermal stability is also required, because the grid is subjected to heating from currents in the element itself, plus heating from the cathode and bombardment of electrons from the cathode. The anode, usually made of copper, conducts the heat of electron bombardment to an external heat sink. Most planar triodes are air-cooled.

Planar triodes designed for operation at 1 GHz and above are used in a variety of circuits. The grounded-grid configuration is most common. The plate-resonant circuit is cavity-based, using waveguide, coaxial line, or stripline. Electrically, the operation of a planar triode is much more complicated at microwave frequencies than at low frequencies. Figure 8-18 compares the elements at work for a grounded-grid amplifier operating at low frequencies (a) and at microwave frequencies (b). The equivalent circuit is made more complicated by:

• Stray inductance and capacitance of the tube elements.
• Effects of the tube contact rings and socket elements.

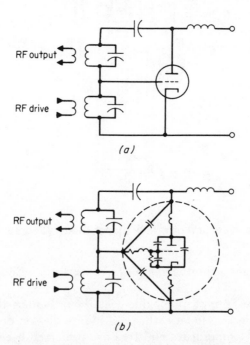

Figure 8-18 Grounded-grid equivalent circuits: (a) low-frequency operation; (b) microwave-frequency operation. The cathode-heating and grid-bias circuits are not shown. (*Source:* Varian Associates.)

- Distributed reactance of cavity resonators and the device itself.
- Electron transit-time effects, which result in resistive loading and phase shifts.

Reasonable gains of 5-10 dB may be achieved with a planar triode. Increased gain is available by cascading stages. Interstage coupling may consist of waveguide or coaxial-line elements. Tuning is accomplished by varying the cavity inductance or capacitance. Additional bandwidth is possible by stagger-tuning cascaded stages.

8.6 Microwave Tube Life

Any analysis of microwave tube life must first identify the parameters that define *life*. The primary *wear out* mechanism in a microwave power tube is the electron gun at the cathode. In principal, the cathode will eventually evaporate the activating material and cease to produce the required output power. Tubes, however, rarely fail because of low emission, but for a variety of other reasons that are usually external to the device.

Power tubes designed for microwave applications provide long life when operated within their designed parameters. The point at which the device fails to produce the required output power can be predicted with some accuracy, based on design data and in-service experience. Most power tubes, however, fail because of mechanisms other than predictable chemical reactions inside the device itself. External forces, such as transient overvoltages caused by lightning, cooling system faults, and improper tuning more often than not lead to the failure of a microwave tube.

8.6.1 Filament Voltage Control

Extending the life of a microwave tube begins with accurate adjustment of filament voltage. The filament should not be operated at a reduced voltage in an effort to extend tube life. Unlike a thoriated tungsten grid tube, reduced filament voltage may cause uneven emission from the surface of the cathode with little or no improvement in cathode life.

Voltage should be applied to the filament for a specified warm-up period before the application of beam current to minimize thermal stress on the cathode/gun structure. However, voltage should not be applied to the filaments for extended periods (2 hours or more) if no beam voltage is present. The net rate of evaporation of emissive material from the cathode surface is greater without beam voltage. Subsequent condensation of material on gun components may lead to voltage standoff problems.

8.6.2   Life-Support System

Transmitter control logic is usually configured for two states of operation:

- *Operational level.* Requires all of the "life-support" systems to be present before the HV command is enabled.
- *Overload level.* Removes HV when one or more fault conditions occur.

The cooling system is the primary life-support element in most transmitters. The cooling system should be fully operational before the application of voltages to the tube. Likewise, a cool-down period is usually recommended between the removal of beam and filament voltages and shut-down of the cooling system.

Most microwave power tubes require a high-voltage removal time of less than 100 ms from the occurrence of an overload. If the trip time is longer, damage may result to the device.

Arc detectors are often installed in the cavities of high power tubes to sense fault conditions, and shut down the high-voltage power supply before damage can be done to the tube. Depending on the circuit parameters, arcs can be sustaining, requiring removal of high voltage to squelch the arc. A number of factors can cause RF arcing, including:

- Overdrive condition
- Mistuning of one or more cavities
- Poor cavity fit (applies to external types only)
- Under-coupling of the output to the load
- Lightning strike at the antenna
- High VSWR

Regardless of the cause, arcing can destroy internal elements or the vacuum seal if drive and/or high voltage are not removed quickly. A lamp is usually included with each arc-detector photocell for test purposes.

8.6.3   Cooling System

The cooling system is vital to any transmitter. In a high-power microwave transmitter, the cooling system may need to dissipate as much as 70 percent of the input ac power in the form of waste heat. For vapor phase-cooled devices, pure (distilled or demineralized) water must be used. Because the collector is usually only several volts above ground potential, it is generally not necessary to use deionized water.

The considerations outlined in Chapter 7 (Section 7.5.4) also apply to cooling microwave power tubes.

8.6.4   Reliability Statistics

Determination of the *mean-time-between-failure* (MTBF) of a microwave tube provides valuable information on the operation of a device in a given installation. The following formulas can be used to predict tube life in operating years, and the number of replacement tubes that will be needed during the life of a transmitter.

$$Y = \frac{MTBF}{H}$$

$$N = \frac{L \times S}{Y}$$

Where:

Y	=	Tube life in operating years
MTBF	=	Tube mean-time-between-failure (gathered from manufacturer's literature or on-site experience)
H	=	Hour of operation per year
	=	365 x hours per operating day
N	=	Number of tubes needed over the life of the transmitter
L	=	Anticipated life of the transmitter in years
S	=	Number of tubes per transmitter
R	=	Number of replacement tubes needed over life of the transmitter

MTTR. The *mean-time-to-repair* (MTTR) of a given piece of RF equipment is an important consideration for any type of facility. MTTR defines the *maintainability* of a system. In the case of a microwave transmitter, the time required to change a tube is an important consideration, especially if standby equipment is not available. The time-change estimates listed previously in this chapter (one hour for an integral cavity klystron and four to six hours for a 60 kW UHF-TV external-cavity tube) assume that no preparation work has been performed on the spare device.

Pre-tuning a tube is one way to shorten the MTTR for an external-cavity device. The spare tube is installed during a maintenance period and the system is tuned for proper operation. After documenting the positions of all tuning controls, the tube is removed and returned to its storage container, along with the list of tuning-control readouts. In this way, the external-cavity tube can be placed in service much faster during an emergency. This procedure will probably not result in a tube tuned for optimum performance, however, it may provide a level of performance that is acceptable on a temporary basis.

Consideration of MTTR is important for a facility because most microwave tubes fail from a mechanism other than reduced-cathode emission, which is a *soft failure* that can be anticipated with some degree of accuracy. Catastrophic failures, on the other hand, offer little — if any — warning.

8.7   Practical Application

Klystrons have been the mainstay of UHF broadcasting for decades. They provide reliable service and high performance. An examination of a typical high power UHF transmitter will help put the principles of microwave power tubes into perspective.

8.7.1   UHF-TV Transmitter

A 60-kW transmitter is shown in block diagram form in Figure 8-19. A single high-power klystron is used in the visual amplifier, and another is used in the aural amplifier. The tubes are driven from solid-state intermediate power amplifier (IPA) modules. The transmitter utilizes ACE-type beam control, requiring additional pre-distortion to compensate for the non-linearities of the final visual stage. Pre-distortion is achieved by correction circuitry at an intermediate frequency (IF) in the modulator. Both klystrons are driven from the output of a circulator, which assures a minimum of driver-to-load mismatch problems.

A block diagram of the beam-modulator circuit is shown in Figure 8-20. The system receives input signals from the modulator, which synchronizes ACE pulses to the visual tube with the video informa-

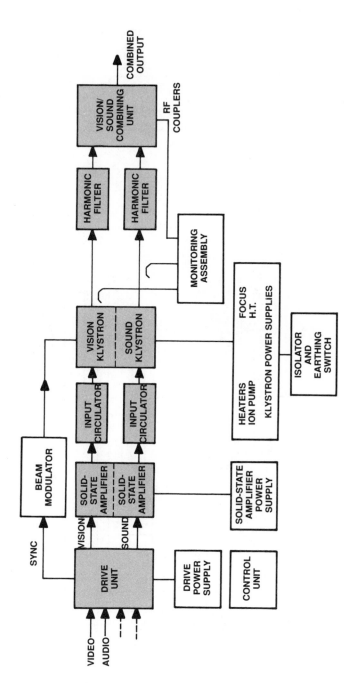

Figure 8-19 Schematic diagram of a 60-kW klystron-based television transmitter. (*Courtesy of Marconi.*)

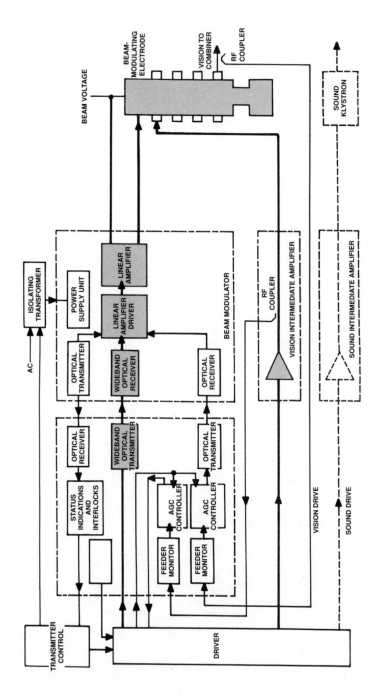

Figure 8-20 Schematic diagram of the modulator section of a 60-kW television transmitter.

tion. The pulse waveform is developed through a pulse amplifier, rather than a switch. This permits more accurate adjustments of operating conditions of the visual amplifier.

Although the current demand from the beam modulator is low, the bias is near cathode potential, which is at a high voltage relative to ground. The modulator, therefore, must be insulated from the chassis. This is accomplished with optical transmitters and receivers connected via fiber optic cables. The fiber optic lines carry supervisory, gain control and modulating signals.

The four-cavity external klystrons will tune to any channel in the UHF-TV band. An adjustable beam perveance feature enables the effective electrical length of the device to be varied by altering the beam voltage as a function of operating frequency. Electromagnetic focusing is used on both tubes. The cavities, body, and gun areas of the klystrons are air-cooled. The collectors are vapor-phase cooled using an external heat-exchanger system. The outputs of the visual and aural klystrons are passed through harmonic filters to an RF combiner before being applied to the antenna system.

Bibliography

1. Fink, D. and D. Christiansen, *Electronics Engineer's Handbook*, Third Edition, McGraw-Hill, New York, 1989.
2. Dick, Bradley, "New Developments in RF Technology," *Broadcast Engineering Magazine*, Intertec Publishing, Overland Park, KS, May 1986.
3. Whitaker, J. and T. Blankenship, "Comparing Integral and External Cavity Klystrons," *Broadcast Engineering*, Intertec Publishing, Overland Park, KS, November 1988.
4. Ridgwell, J. F., "A New Range of Beam-Modulated High-Power UHF Television Transmitters," *Communications & Broadcasting*, Number 28, Marconi Communications, Clemsford, Essex, England, 1988.
5. *IEEE Standard Dictonary of Electrical and Electronics Terms*, Institute of Electrical and Electronics Engineers, New York, 1984.

9

New Developments in Microwave Tube Technology

9.1 Introduction

The operating efficiency of a high-power transmitter is important to every end-user. The penalties for low efficiency include high operating costs, shortened tube life, and increased complexity of the cooling system. Much work is currently underway to improve the efficiency of power tubes used in UHF-TV broadcasting. While current applications center around television transmission, the technologies being developed may also be applied to many other uses.

Improvements in power device technology have made it possible to dramatically reduce the cost of operating a UHF television transmitter. Improvements in ac-to-RF efficiency continue to be made as new devices come into production, and as new tuning and pulsing techniques are perfected. The driving force behind this work has been economics. UHF broadcasters utilize high transmitting power in order to provide adequate coverage to their service area, and that high power costs money.

The workhorse of UHF broadcasting today is the klystron. Much effort has been directed toward making the klystron a more efficient device. Parallel development has also occurred for tetrode-based systems, which offer good efficiency and straightforward design. Power levels of 25 kW and more are now practical with tetrodes. For higher powers, the *Klystrode tube* and the *multi-stage depressed collector* (MSDC) klystron show great promise. Although totally different designs, the two devices provide essentially the same end result: a 200 percent efficiency improvement over conventional klystrons.

9.1.1   Comparing Efficiency

Comparing the efficiency figures of television transmitters is complicated by the many variables involved. Any examination of efficiency must be tempered with an understanding of the measurement parameters. Some manufacturers specify overall transmitter ac-to-RF efficiency, including the cooling system. This number is really what the end user is concerned with. With klystron-based transmitters, the efficiency of the final amplifying stage is also important because that is where most of the energy is expended. Because the klystron is a class A device, the average dc-input power does not vary significantly with picture content. The *figure-of-merit* (FOM) is defined as:

$$\text{Figure of merit} = \frac{\text{RF Peak Power Output}}{\text{Average dc Input at 50\% APL}}$$

Note: APL = *Average picture level* (a reference video image used in the broadcast industry as a measure of comparison)

Achievements over the last two decades to ease the burden faced by UHF stations have been evolutionary. Early klystrons for television service had a FOM of between 0.30 and 0.40. The introduction of mod-anode pulsing enabled FOM performance of greater than 0.40 to be achieved using these tubes. Improved designs and new methods of tuning, which traded gain for efficiency, brought the basic-tube FOM to more than 0.40. External-cavity klystrons of this efficiency that were fitted with a pulsed ACE electrode further improved the FOM to between 0.50 and 0.60. The latest generation of external-cavity klystrons has achieved a basic FOM of 0.50, which when pulsed may be raised to between 0.60 and 0.75. Integral-cavity ACE tubes are now in service with an FOM in excess of 0.80.

A high-efficiency klystron, fully pulsed and tuned with full linearity compensation, may be expected (under ideal conditions) to achieve a figure of merit 1.69 times its out-of-the-box performance.[1] Given a tube that has a basic efficiency of 50 percent, a pulsed peak-of-sync efficiency of 84.5 percent is the best that can be achieved.

---

1   Ostroff, N., A. Whiteside, and L. Howard, "An Integrated Exciter-Pulser System for Ultra High Efficiency Klystron Operation," *Proceedings of the NAB Engineering Conference*, Las Vegas, 1985.

These levels are not, however, seen in actual broadcast operation because fully pulsed linearity correction is difficult to achieve on a stable basis. Practical values of 60 percent to 70 percent are common, although values as high as 80 percent have been reported.

The next step up the efficiency ladder is revolutionary, not evolutionary. The Klystrode tube, the first true high-efficiency high-power UHF transmitting device to go into regular service, and the MSDC klystron achieve a dramatic leap in FOM.

9.2 Klystrode Tube

The Klystrode tube is the end result of a development program started (at Varian Associates) in 1980 with UHF-TV in mind as a major application. By 1982, test results on an early version were encouraging enough to justify the publication of a an IEEE paper with the title, "The Klystrode — An Unusual Transmitting Tube With Potential for UHF-TV."[2] Its full potential was not realized until a great deal more development had been done, including major improvements to both the tube and its associated RF circuitry.

The basic concept of the Klystrode dates back to the late 1930s, but it was only within the last decade that serious engineering effort was put into the tube to make it a viable product for high-power UHF service. The fundamental advantage of the Klystrode is its ability to operate at class B level. The result is higher efficiency when compared to a conventional klystron.

9.2.1 Theory of Operation

As its name implies, the Klystrode tube is a hybrid between a klystron and a tetrode. The high reliability and power-handling capability of the klystron is due, in part, to the fact that electron beam dissipation takes place in the collector electrode, which is separate from the RF circuitry. The electron dissipation in a tetrode is at the anode and the screen grid, both of which are an inherent part of the RF circuit and must, therefore, be physically small at UHF frequencies. The tetrode, on the other hand, has the advantage that modulation is produced directly at the cathode by a grid, so that a long drift

---

2 Priest, D. and M. Shrader, "The Klystrode — An Unusual Transmitting Tube With Potential for UHF-TV," Proceedings of the IEEE, Volume 70, Number 11, New York, November 1982.

space is not required to produce density modulation. The Klystrode tube has a similar advantage over the klystron — high efficiency in a small package.

The Klystrode tube is shown schematically in Figure 9-1. The electron beam is formed at the cathode, density modulated with the input RF signals by a grid, and then accelerated through the anode aperture. In its bunched form, the beam drifts through a field-free region and then interacts with the RF field in the output cavity. Power is extracted from the beam in the same way as a klystron. The input circuit resembles a typical UHF power grid tube. The output circuit and collector resemble a klystron.

A production version of the 60-kW device is shown in Figure 9-2. Double-tuned cavities are used to obtain the required bandwidth. The load is coupled at the second cavity (Figure 9-3). This arrangement has proven to be an attractive way to couple power out of the device because no coupling loop or probe is required in the primary cavity, a problem at the high end of the UHF band.

The potential for failure of the tube's pyrolytic graphite grid has received a good deal of attention from designers. Protection of the grid begins with the basic tube geometry. The grid is placed in a location away from potential arc paths. Protection external to the tube is centered around a fast, high-current crowbar system that limits the energy that can be delivered to transmitter components during an arc or other fault. Two tubes cover the UHF-TV band, with the dividing point at Channel 35 (602 MHz). The low band version of the tube differs only in the height of the output cavities and

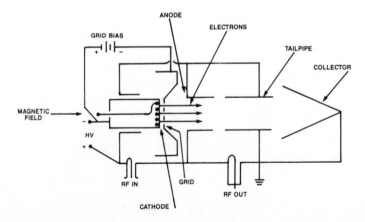

Figure 9-1 Simplified schematic diagram of the Klystrode tube. (*Source: Broadcast Engineering Magazine.*)

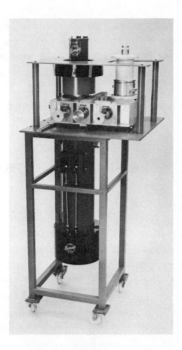

Figure 9-2 The 60-kW Klystrode tube mounted in its support stand with the output cavity attached. (*Source: Broadcast Engineering Magazine.*)

the length of the input circuit. Performance measurements on the low-band tube are similar to the high-band version and indicate good bandwidth, efficiency and power gain.

Figure 9-3 A close-up view of the double-tuned output cavity of the 60-kW Klystrode tube. (*Source: Broadcast Engineering Magazine.*)

Because the Klystrode tube provides both beam-power variation during sync pulses (as in a pulsed klystron) and variation of beam power over the active video waveform, it is capable of high efficiency. The device provides full-time beam modulation as a result of its inherent structure and class B operation. The FOM for a Klystrode tube has been consistently measured at 1.20 or higher.

9.2.2   Applying the Klystrode

The first Klystrode tube to be placed into service was at a UHF-TV station in Wrens, Georgia, (WCES) in 1988. The project successfully tested the concept. A block diagram of the 120-kW transmitter is shown in Figure 9-4. Reliability was an overriding consideration, with as many as four *fall-back* operating positions designed into the transmitter and combiner. The crowbar circuitry was designed to protect the pyrolytic graphite grid during an arc condition inside the tube. As it turned out, the grid is not damaged by an arc without the crowbar, but the station can get the Klystrode back on the air much faster because less gas is generated in the tube with the protection circuit. A block diagram of the crowbar is shown in Figure 9-5. The trip point is set at 50 A. The response time of the crowbar is less than 10 ms. The peak current permissible through the discharge tube is 3,000 A.

Because of the Klystrode's class B operation, the response of the power supply to a varying load is an important design parameter. If

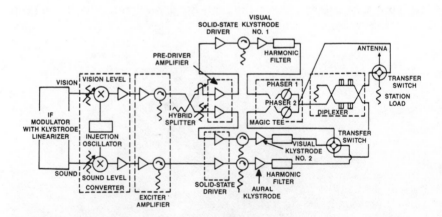

Figure 9-4 Block diagram of the 120-kW (Comark) Klystrode transmitter installed at WCES in Wrens, GA. Redundancy was a primary concern in the design of the transmission system. (*Source: Broadcast Engineering Magazine.*)

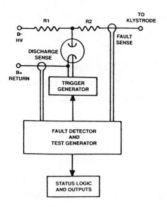

Figure 9-5 Simplified block diagram of the crowbar circuit designed into the Klystrode SK-series (Comark) transmitters. The circuit is intended to protect the tube from potentially damaging fault currents. (*Source: Broadcast Engineering Magazine.*)

a standard klystron beam supply were used in a Klystrode-equipped transmitter, performance under varying picture levels would be unacceptable. Consider the application of a *bounce* (black-to-white) signal to the transmitter. Beam current for a typical Klystrode 60-kW transmitter would change from approximately 400 mA to 2 A. The effect with a conventional klystron supply would be ringing of about 20 percent on the beam voltage. The Klystrode power supply must, therefore, be designed for tight transient regulation. The transmitter is set up as a closed-loop system. The output signal, after the combiner, is sampled at the pedestal level and used as a reference to control the IF drive going to the driver amplifiers. The AGC loop corrects for long-term sags in beam voltage caused by changes in utility company input power.

The Klystrode tube does not have the same hard saturation characteristic as the klystron. The Klystrode high-power transfer curve flattens more slowly than a klystron and continues to increase with increasing drive power, requiring a different approach to linearity correction. Correction is required at white level, black, and sync. Further, correction must be done at IF after the SAW filter to permit proper sideband cancellation. It is possible to improve linearity by increasing the tube idling current. This would, however, degrade the overall efficiency of the system. Extended linearity correction is, instead, built into the modulator. The system independently corrects:

* Low-frequency linearity
* Differential gain
* Differential phase

- ICPM
- In-band response flatness

The modulator includes the capability for sync reinsertion and transmitter output power control. The power AGC system uses a sample of the transmitter output signal for comparison against a preset power reference.

9.2.3   Test Data

Test data on the Klystrode transmitter shows exceptional performance. Of particular interest is the lower sideband regeneration. Figure 9-6 shows lower sideband response of the transmitter operating at rated power into a dummy load. Lower sideband regeneration

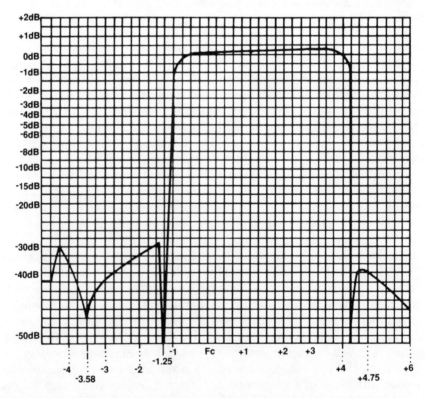

Figure 9-6 Measured lower sideband regeneration of the WCES Klystrode transmitter taken at full operating power into a dummy load. Note that the lower sideband is held to a maximum of -30 dB. (*Source: Broadcast Engineering Magazine.*)

is a measure of the performance of the overall transmitter. Ideally, the lower sideband should be as low out of the final amplifier as it is out of the modulator (approximately -40 dB). Because of non-linearities in the power amplifier, however, some lower sideband energy is usually regenerated through intermodulation.

The FCC spec for lower sideband regeneration is -20 dB. The limiting factor in most klystron transmitters, and the factor that establishes a break-even point on pulsing, is the lower sideband regeneration level. As the klystron is pulsed deeper, more non-linearity results. To compensate, more correction is needed and at some point the non-linearity exceeds the range of correction available. The worst case lower sideband reading of the first Klystrode transmitter placed into service was -30 dB. That measurement is better than the typical performance of a multiplexed (common aural/visual amplification) transmitter, which is designed specifically to have low intermodulation. Figure 9-7 shows low-frequency linearity of the transmitter, taken with a stair-step waveform. Linearity is within about one percent, confirming the lower sideband regeneration data.

Waveform monitor and vectorscope displays with color bars are shown in Figure 9-8 and Figure 9-9, respectively. These traces, and other data taken after installation, show performance, essentially identical to a conventional klystron-based transmitter.

Operating Efficiency. The WCES transmitter and the other installations that have followed it all easily met FCC proof-of-performance specifications. The most interesting parameter — and the one

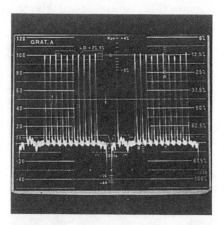

Figure 9-7 Low-frequency linearity performance of the WCES transmitter. (Scope photos courtesy of Georgia Public TV and Comark. *Source: Broadcast Engineering Magazine.*)

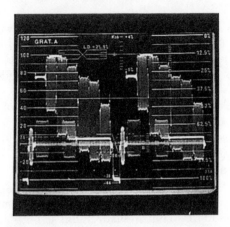

Figure 9-8 Waveform monitor display of color bars (horizontal rate) taken during the WCES-TV proof of performance. (*Source: Broadcast Engineering Magazine.*)

of most importance to TV engineers — is efficiency. To use WCES as an example, the average beam current at 50 percent APL and 104 percent peak RF output was 1.65 A per tube. At black level the beam current was 2.31 A per tube. Combined with a beam voltage of 31.5 kV, the measured FOM was 1.20 and the total beam power was 104 kW. As a means of comparison, for a fully pulsed advanced design klystron transmitter, the power consumption of the output visual klystrons at 50 percent APL would have been at least 192 kW. This is based on an assumed 46-percent-efficient klystron using pulsing to

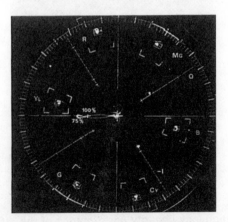

Figure 9-9 Vectorscope display of color bars taken during the WCES-TV proof. (*Source: Broadcast Engineering Magazine.*)

raise the peak-of-sync efficiency to 65 percent. Higher efficiencies can be achieved, but they may not be stable over long-term operation.

It is interesting to note that the measured FOM for the second transmitter installed (WABW) was 1.31. While this represents an 11 percent improvement over the WCES system, the operating cost savings as a result of this extra measure of efficiency were minimal. Figure of merit performance above 1.20 does not result in significant additional power savings because of the law of diminishing returns. At approximately 1.20 FOM, the support circuitry power requirements (driver, power supply losses, and cooling system) begin to have a significant effect on overall transmitter efficiency.

9.2.4 Continuing Research Efforts

Several variations on the basic Klystrode theme have been developed, including air-cooled klystrons operating at 15 kW and 30 kW. Air cooling is practical at these power levels because of the improved efficiency that class B operation provides. Research is also being conducted as of this writing to extend the operating power of the Klystrode tube to 500 kW or more. Designed for scientific research applications, the Klystrode offers numerous benefits over conventional klystron technology. The Klystrode is much smaller than a klystron of similar power, and requires less support circuitry. Because of the improved efficiency, power supply requirements are reduced and device-cooling is simplified.

9.3 MSDC Klystron

Developmental work on the *multi-stage depressed collector* (MSDC) klystron began in the mid 1980s. The project (a joint effort of NASA, several transmitter manufacturers, Varian Associates, and others) has produced a working tube that is capable of efficiency in UHF-TV service that was impossible with previous klystron-based technology.

The MSDC device has potential for both broadcast and non-broadcast applications. NASA originally became involved in the project as a way to improve the efficiency of satellite transmitters. With limited power available on board a space vehicle, efficient operation is critically important. Such transmitters traditionally operate in a linear, non-efficient mode, as do UHF-TV rigs.

Experimentation with depressed collector klystrons dates back to at least the early 1960s. Early products offered a moderate improvement in efficiency, but at the price of greater mechanical and electrical complexity. The MSDC design, while mechanically complex, offers a significant gain in efficiency.

9.3.1   Theory of Operation

MSDC tubes have been built around both integral-cavity and external-cavity klystrons. The devices are essentially identical to a standard klystron, except for the collector assembly. Mathematical models provided researchers with detailed information on the interactions of electrons in the collector region. Computer modeling also provided the basis for optimization of a *beam reconditioning* scheme incorporated into the device. Beam reconditioning is achieved by including a *transition region* between the RF interaction circuit and the collector under the influence of a magnetic field. It is interesting to note that the mathematical models made for the MSDC project translated well into practice when the actual device was constructed.

From the electrical standpoint, the more stages of a multi-stage depressed collector klystron, the better. The tradeoff, predictably, is increased complexity and, therefore, increased cost for the product. There is also a point of diminishing returns that is reached as additional stages are added to the depressed collector system. A four-stage device was chosen because of these factors. As additional stages are added, the resulting improvement in efficiency is proportionally smaller.

Figure 9-10 shows the mechanical configuration of the four-stage MSDC klystron. Note the "V" shape, found through computer studies to provide the best "capture" performance, minimizing electron feedback. A photograph of a partially assembled MSDC collector assembly is shown in Figure 9-11.

Electron Trajectories. The dispersion of electrons in the multi-stage collector is the key element in recovering power from the beam and returning it to the power supply. This is the mechanism that permits greater operating efficiency from the MSDC device. Figure 9-12 illustrates the dispersion of electrons in the collector region during a carrier-only operating mode. Note that there is little dispersion of electrons between stages of the MSDC. Most are attracted to electrode 4, the element at the lowest potential (6.125 kV), referenced to the cathode.

Figure 9-13 shows collector electron trajectories at 25 percent saturation. The electrons exhibit predictable dispersion characteristics during the application of modulation, which varies the velocity of the electrons. This waveform is a reasonable approximation of *average*

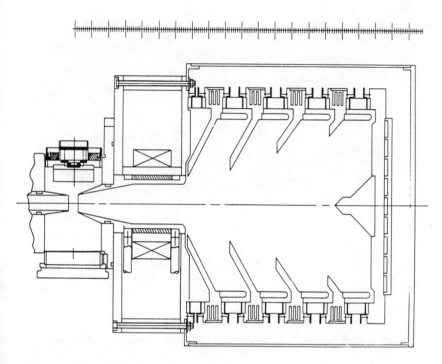

Figure 9-10 Mechanical design of the multi-stage depressed collector assembly. Note the "V" shape of the 4-element system. (*Source: Broadcast Engineering Magazine.*)

modulation for a typical TV image. Figure 9-14 shows electron trajectories at 50 percent saturation, approximately the *blanking* level. Note the increased number of electrons attracted to electrodes 2 and 3, the higher potential electrodes (referenced to the cathode). Figure 9-15 illustrates electron dispersion at 90 percent saturation, approximately the level of sync. As modulation is increased, increasing numbers of electrons are attracted to the higher voltage electrodes. The dramatic increase in electron capture by electrode 1, the highest potential element at a voltage of 24.5 kV, can be observed (referenced to the cathode).

The electrons, thus, sort themselves out in a predictable manner. Notice the arc that is present on many electron traces. The electrons penetrate the electrostatic field of the collector, and are then pulled back to their respective potential.

Power savings are realized because the electrostatic forces set up in the MSDC device slow down the electrons before they come in contact with the copper electrode. The heat that would be produced

Figure 9-11 A partially assembled MSDC collector. *(Courtesy of Varian Associates.)*

in the collector is, instead, returned to the power supply in the form of electrical energy. In theory, peak efficiency would occur if the electrons were slowed down to zero velocity. In practice, however, that is not possible.

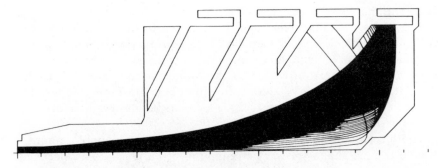

Figure 9-12 Collector electron trajectories for the carrier-only condition. Note that nearly all electrons travel to the last electrode (4), producing electrode current I_4. *(Source: Broadcast Engineering Magazine.)*

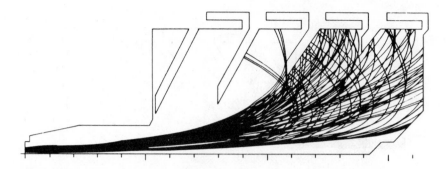

Figure 9-13 Collector trajectories with 25 percent saturation. With the application of modulation, the electrons begin to sort themselves out. (*Source: Broadcast Engineering Magazine.*)

Figure 9-16 shows the distribution of collector current as a function of drive power. With no RF drive, essentially all current goes to electrode 4, but as drive is increased, I_4 drops rapidly as collector current is distributed between the other elements. Note that the current to electrode 5 (cathode potential) peaks at about ten percent of beam current. According to researchers, this suggests that the *secondary yield* of the collector surfaces — a concern addressed during the design stage of the MSDC tube — is within acceptable limits.

Inserted between the klystron and the collector assembly is a refocusing electromagnet that controls the electron beam as it enters the collector region.

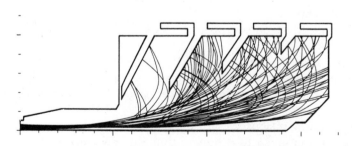

Figure 9-14 Collector electron trajectories at 50 percent saturation, approximately the blanking level. The last three electrodes (2, 3, and 4) share electrons in a predictable manner, producing currents I_2, I_3, and I_4. (*Source: Broadcast Engineering Magazine.*)

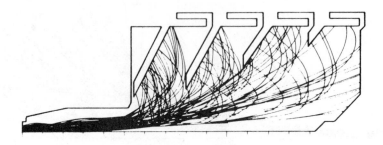

Figure 9-15 Collector trajectories at 90 percent saturation, the sync level. Note the significant increase in the number of electrons attracted to the first electrode, producing I_1. (*Source: Broadcast Engineering Magazine.*)

9.3.2   Mechanical Construction

The completed MSDC assembly is shown in Figure 9-17, with the collector shield partially removed to allow visibility of the collector elements. The collector of the 4-stage MSDC design is actually composed of five elements mounted between ceramic rings for electrical insulation. The fifth electrode is at cathode potential.

Cooling for the MSDC is, not surprisingly, more complicated than a conventional klystron device. The tradeoff, however, is that there is less heat to remove because of the higher efficiency of the device. Water cooling is provided on each electrode of the MSDC tube.

Researchers believe that the MSDC design will have little, if any, effect on the lifetime of the klystron. The electron beam is essentially unchanged. The tube is identical to a conventional integral- or external-cavity klystron except in collector design.

Problem Areas. Preventing feedback within the MSDC device was cited as a potential problem area for designers. Feedback would occur if electrons in the collector, not attracted by any of the electrodes, were to return to the drift tube area of the device. Such an occurrence would seriously distort the linearity of the tube. Particular attention was given to ensuring that the mechanisms which lead to feedback within the MSDC tube did not occur. Other areas of concern involved suppression of secondary electrons, collector cooling and RF radiation.

Suppression of secondary electrons is accomplished both through the mechanical design of the collector and through the use of special

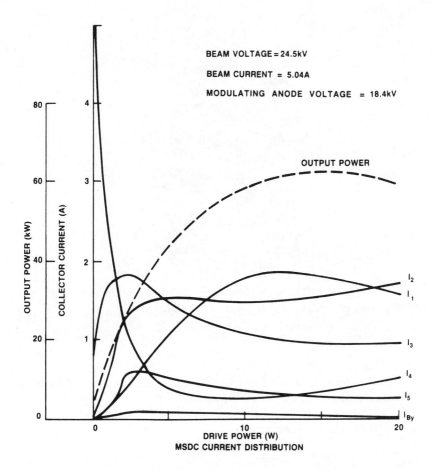

Figure 9-16 The distribution of electrode current as a function of drive power. Note the significant drop in I_4 as drive power is increased. I_5 is the electrode at cathode potential. (*Source: Broadcast Engineering Magazine.*)

materials on the collector surfaces. Materials are available that exhibit low secondary yield, such as carbon. But carbon is known to absorb gases, a potentially serious problem in a vacuum tube environment. The solution involves applying only enough carbon to keep secondary electrons at a low level. If the carbon coating on the collector assembly is kept thin, only a small volume is available to absorb gases.

Another area of concern was the potential for RF radiation. By its very nature, video currents exist in the elements of the collector, and the design of the power supply must take that into consideration.

Figure 9-17 The MSDC collector assembly with the protective shield partially removed. *(Courtesy of Varian Associates.)*

Interestingly, it was found that the main problem relating to radiation would probably come from the video component of the klystron waveform, rather than the RF component.

9.3.3   MSDC Power Supply

Design criteria for the collector power supply system provides a mixed bag of requirements. The critical parameter is the degree of regulation between the cathode and anode. The relative differences between the elements of the collector do appear to be significant. Consequently, the bulk of the power supplied to the tube does not need to be well-regulated. This is in contrast with current klystron operation, in which the entire beam power supply must be regulated. This factor effectively decreases the amount of power that needs to be regulated to 1 to 2 percent of the dc input, offsetting to some extent the additional cost involved in constructing multiple supplies to facilitate the 4-stage MSDC design. Two approaches can be taken to collector supply design: Figure 9-18 shows a power supply using parallel arrangement of the power units, and Figure 9-19 shows a series-constructed system. Note that in both cases, the collector electrodes are stepped at a 6.125 kV potential difference for each element.

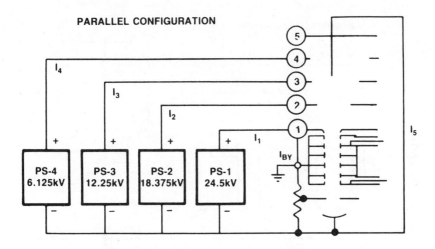

Figure 9-18 Parallel configuration of the MSDC power supply. Each power supply section has an output voltage that is an integral multiple of 6.125 kV. (*Source: Broadcast Engineering Magazine.*)

9.3.4   Device Performance

The efficiency improvement of the MSDC klystron over a conventional klystron is impressive. Efficiency measurements on prototype 60-kW MSDC tubes show a typical FOM of 1.32 when pulsed. This represents an efficiency of more than twice a conventional klystron.

The bandpass performance of the device is another critical parameter. Figure 9-20 charts power output as a function of frequency and RF drive at full power. (Drive power is charted from 0.5 W to 16.0 W). Figure 9-21 charts power output as a function of frequency and RF drive with the tube in a beam pulsing mode. (Drive power is charted from 0.5 W to 32.0 W). Note that the traces provide good linearity over a 6 MHz bandwidth. Gain, as a function of frequency and power, is essentially constant and undisturbed.

9.3.5   Applying the MSDC Klystron

As of this writing, two MSDC transmitters had been placed into on-air service. A block diagram of one system (Varian TVT) is shown in Figure 9-22. The 60-kW transmitter incorporates two external-cavity MSDC klystrons, one for the visual and another for the aural. Design of the transmitter is basically identical to a non-MSDC system,

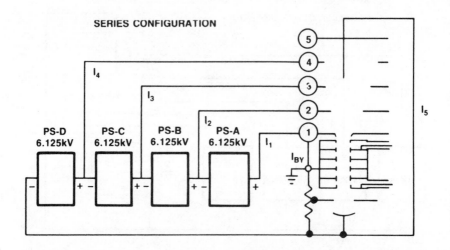

Figure 9-19 Series configuration of the MSDC power supply. This arrangement uses 4 identical 6.125-kV power supplies connected in series to achieve the needed voltages. (*Source: Broadcast Engineering Magazine.*)

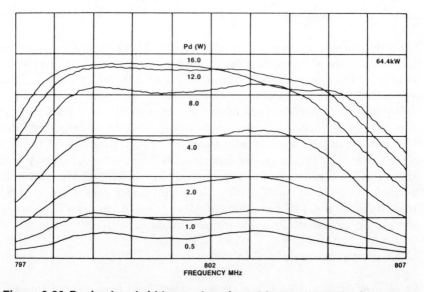

Figure 9-20 Device bandwidth as a function of frequency and drive power. Beam voltage is 24.5 kV and beam current is 5.04 A for an output power of 64 kW. These traces represent the full power test of the MSDC device. (*Source: Broadcast Engineering Magazine.*)

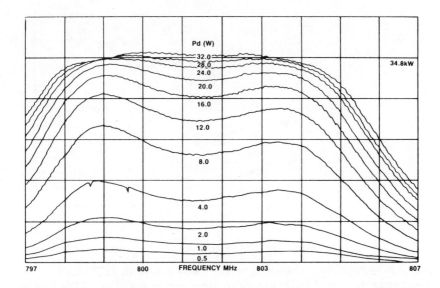

Figure 9-21 Device bandwidth in the beam-pulsing mode as a function of frequency and drive power. Beam voltage is 24.5 kV and beam current is 3.56 A with an output power of 34.8 kW. (*Source: Broadcast Engineering Magazine.*)

with the exception of the power supply and cooling system. The efficiency available from the MSDC makes further device improvements subject to the law of diminishing returns, as in the Klystrode tube system. Support equipment begins to consume an increasingly large share of the power budget as output device efficiency is improved. Tuning of the MSDC klystron is the same as a conventional klystron, and the same magnetic circuit is used.

The power supply arrangement for the example MSDC system is shown in Figure 9-23. A series beam supply was chosen for technical and economic reasons. Although the current to each collector varies widely with instantaneous output level, the total current stays within narrow limits. Using a single transformer for all supplies, therefore, minimizes the size of the iron core required. A 12-pulse rectifier bank provides low ripple and reduces the need for additional filtering. The size and complexity of the rectifier stack is increased little beyond a normal beam supply, because the total potential of the four supplies is similar to a normal klystron transmitter (24.5-27.5 kV).

The collector stages of the MSDC device are water-cooled by a single water path that loops through each electrode element. Because

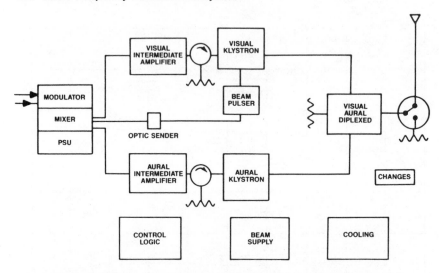

Figure 9-22 Simplified block diagram of a 60-kW (Varian TVT) MSDC transmitter. The aural klystron may utilize a conventional or MSDC tube at the discretion of the user. *(Courtesy of Varian TVT.)*

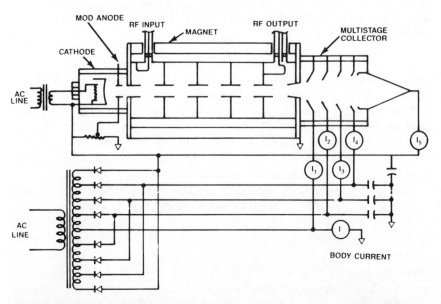

Figure 9-23 Simplified schematic diagram of the power supply for an MSDC klystron transmitter. A single high voltage transformer with multiple taps is used to provide the needed collector voltage potentials. *(Courtesy of Varian TVT.)*

high voltage is present on individual elements, purity of the water is critical to proper operation. A two-stage system is used with a water-to-water plate heat exchanger separating the primary and secondary systems.

9.4 High Power Tetrode

New advancements in vacuum-tube technology have permitted the construction of high power UHF transmitters based on tetrodes. Such devices are attractive because they inherently operate in an efficient class A-B mode. At least one 25-kW tetrode-based television transmitter (Acrodyne) is now on the air in common amplification service. The TH563 tube (Thomson) is capable of 35 kW peak-of-sync power output in split (separate aural/visual) operation.

UHF tetrodes operating at high-power levels provide essentially the same specifications, gain and efficiency as tubes operating at lower powers. The anode power supply is much lower in voltage than the collector potential of a klystron- or Klystrode tube-based system (8 kV is common). The tetrode also does not require focusing magnets.

Efficient removal of heat is the key to making a tetrode practical at high-power levels. The TH563 uses water-cooling. Air-cooling at such levels is impractical because of the fin size that would be required. Also, the blower for the tube would have to be quite large, reducing the overall transmitter ac-to-RF efficiency.

The expected lifetime of a tetrode in UHF service is shorter than a klystron of the same power level. Typical lifetimes of 8,000 to 15,000 hours have been reported. It must be noted, however, that the replacement cost of a tetrode is much less than a klystron or Klystrode tube.

Work is underway on methods to extend the operating limits of the tetrode, while still retaining the benefits of its inherent class A-B operation. Tetrodes designed for 50-kW peak-of-sync power have been considered by at least one tube manufacturer.

Bibliography

1. Ostroff, N., A. Whiteside, and L. Howard, "An Integrated Exciter/Pulser System for Ultra High-Efficiency Klystron Operation," *Proceedings of the NAB Engineering Conference,* Las Vegas, 1985.
2. Shrader, Merrald B., "Klystrode Technology Update," *Proceedings of the NAB Engineering Conference,* Las Vegas, 1988.

3. Ostroff, N., A. Whiteside, A. See, and R. Kiesel, "A 120 kW Klystrode Transmitter for Full Broadcast Service," *Proceedings of the NAB Engineering Conference,* Las Vegas, 1988.
4. Priest, D. and M. Shrader, "The Klystrode — an Unusual Transmitting Tube with Potential for UHF-TV," *Proceedings of the IEEE,* Volume 70, Number 11, New York, November 1982.
5. Badger, George, "The Klystrode: A New High-Efficiency UHF-TV Power Amplifier," *Proceedings of the NAB Engineering Conference,* Dallas, 1986.
6. Ostroff, N., R. Kiesel, A. Whiteside, and A. See, "Klystrode-Equipped UHF-TV Transmitters: Report on the Initial Full Service Station Installations," *Proceedings of the NAB Engineering Conference,* Las Vegas, 1989.
7. Technical background information on the Vista line of transmitters, Varian TVT, Cambridge, England, 1989.
8. McCune, Earl, "Final Report: The Multi-Stage Depressed Collector Project," *Proceedings of the NAB Engineering Conference,* Las Vegas, 1988.
9. Technical background information on UHF tetrode operation, Dr. Timothy Hulick, Acrodyne, Blue Bell, PA, 1988.

10

Transmission Line and Waveguide

10.1 Introduction

The mechanical and electrical characteristics of the transmission line, waveguide and associated hardware that carries power from the transmitter to the antenna are critical to the proper operation of any RF system. Mechanical considerations determine the ability of the components to withstand temperature extremes, lightning, rain and wind. In other words, they determine the overall reliability of the system. A number of different types of hardware are available to users. The approach taken depends on the power level, frequency, length of the run from the transmitter to the antenna and the installation method preferred.

10.2 Coaxial Transmission Line

Two types of coaxial transmission lines are in common use today: *rigid* line and *corrugated* (*semi-flexible*) line. Rigid coaxial cable is constructed of heavy-wall copper tubes with Teflon or ceramic spacers. (Teflon is a registered trademark of DuPont.) Rigid line provides electrical performance approaching an ideal transmission line, including: high-power handling capability, low loss, and low VSWR (voltage standing wave ratio). Rigid transmission line is, however, expensive to purchase and install.

The primary alternative to rigid coax is semi-flexible transmission line made of corrugated outer- and inner-conductor tubes with a spiral polyethylene or Teflon insulator (see Figure 10-1). Semi-flexible line has three primary benefits:

- It is manufactured in a continuous length, rather than 20-ft. sections as typical for rigid line.
- Because of the corrugated construction, the line may be shaped as required for routing from the transmitter to the antenna.
- The corrugated construction permits differential expansion of the outer and inner conductors.

Each size of line has a minimum bending radius. For most installations, the flexible nature of corrugated line permits the use of a single piece of cable from the transmitter to the antenna, with no

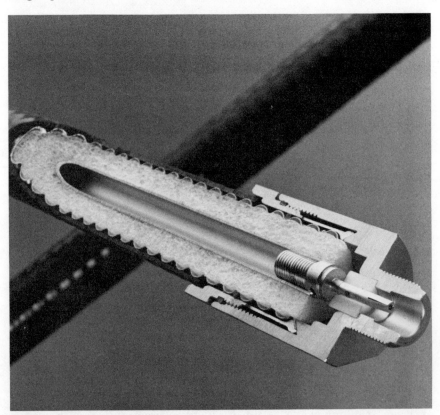

Figure 10-1 Cross-section of a length of semi-flexible coaxial cable with a type *N* connector attached. *(Courtesy of Andrew Corp.)*

elbows or other transition elements. This speeds installation and provides for a more reliable system.

10.2.1   Theory of Operation

A coaxial transmission line consists of concentric center and outer conductors that are separated by a dielectric material. When current flows along the center conductor, it establishes an electric field. The electric flux density and the electric field intensity are determined by the dielectric constant of the dielectric material. The dielectric material becomes polarized with positive charges on one side and negative charges on the opposite side. The dielectric, therefore, acts as a capacitor with a given capacitance per unit length of line. Properties of the field also establish a given inductance per unit length, and a given series resistance per unit length. If the transmission line resistance is negligible and the line is terminated properly, the following formula describes the characteristic impedance (Z0) of the cable:

$$Z_o = \sqrt{L/C}$$

Where:
L = inductance in H/ft
C = capacitance in F/ft

Coaxial cables typically are manufactured with 50 Ω or 75 Ω characteristic impedances. Other characteristic impedances are possible by changing the spacing between the center and outer conductors.

Velocity of Propagation. A signal traveling in free space is unimpeded and has a free-space velocity equal to the speed of light. In a transmission line, capacitance and inductance slow the signal as it propagates along the line. The amount the signal is slowed is represented as a percentage of the free-space velocity. This quantity is called the *relative velocity of propagation* and is described by the equation:

$$V_p = \frac{1}{\sqrt{L \times C}}$$

Where:
L = inductance in H/ft
C = capacitance in F/ft

and
$$V_r = (V_p/c) \times 100 \text{ percent}$$

Where:
V_p = velocity of propagation
$c = 9.842 \times 10^8$ ft./s (free-space velocity)
V_r = velocity of propagation as a percentage of free-space velocity

Transverse Electromagnetic Mode. The principal mode of propagation in a coaxial line is the *transverse electromagnetic mode* (TEM). This mode will not propagate in a waveguide, and that is why coaxial lines can propagate a broad band of frequencies efficiently. The cut-off frequency for a coaxial transmission line is determined by the line's dimensions. Above cut-off, modes other than TEM can exist and the transmission line properties are no longer defined. The cut-off frequency is equivalent to:

$$f_c = \frac{7.50 \times V_r}{D_i + D_o}$$

Where:
f_c = cut-off frequency in GHz
V_r = velocity (percent)
D_i = inner diameter of outer conductor in inches
D_o = outer diameter of inner conductor in inches

At dc, current in a conductor flows with uniform density over the cross section of the conductor. At high frequencies, the current is displaced to the conductor surface. The effective cross section of the conductor decreases and the conductor resistance increases. For RF signals, current only flows in a thin "skin" of the conductor (the *skin effect*).

Center conductors are made from copper-clad aluminum or high-purity copper and can be solid, hollow tubular or corrugated tubular. Solid center conductors are found on semi-flexible cable with 1/2-in. diameter or smaller. Tubular conductors are found in 7/8-in. diameter or larger cables. Although the tubular center conductor is used primarily to maintain flexibility, it can also be used to pressurize the antenna through the feeder.

Dielectric. Coaxial lines use two types of dielectric construction to isolate the inner conductor from the outer conductor. The first is an air dielectric, with the inner conductor supported by a dielectric spacer and the remaining volume filled with air or nitrogen gas. The spacer, which may be constructed of spiral or discrete rings, is typically made of Teflon or polyethylene. Air-dielectric cable offers lower attenuation and higher average power ratings than foam-filled cable, but requires pressurization to prevent moisture entry.

Foam-dielectric cables are ideal for use as feeders with antennas that do not require pressurization. The center conductor is surrounded completely by foam-dielectric material, resulting in a high dielectric-breakdown level. The dielectric materials are polyethylene-based formations, which contain anti-oxidants that reduce dielectric deterioration at high temperatures.

10.2.2   Electrical Considerations

Voltage standing wave radio (VSWR), attenuation, and power handling capability are key electrical factors in the application of coaxial cable. High VSWR can cause power loss, voltage breakdown, and thermal degradation of the line. High attenuation means less power delivered to the antenna, higher power consumption at the transmitter, and increased heating of the transmission line itself.

VSWR is a common measure of the quality of a coaxial cable. High VSWR indicates non-uniformities in the cable that can be caused by variations in the dielectric core diameter, variations in the outer conductor, poor concentricity of the inner conductor, or a non-homogeneous or periodic dielectric core. Although each of these may contribute only a small reflection, they can add up to a measurable VSWR at a particular frequency.

Rigid transmission line is typically available in a standard length of 20 ft., and in alternative lengths of 19.5 ft. and 19.75 ft. The shorter lines are used to avoid VSWR buildup, caused by discontinuities resulting from the physical spacing between line section joints. If the section length and the operating frequency have a half-wave correlation, connector junction discontinuties will add. This effect is known as *flange build-up*. The result can be excessive VSWR. The critical frequency at which a half-wave relationship exists is given by:

$$F_{cr} = \frac{490.4 \times n}{L}$$

Where:
F_{cr} = the critical frequency
n = any integer
L = transmission line length in feet

Table 10-1 Representative specifications for various types of flexible air-dielectric coaxial cable. (*Source: Broadcast Engineering Magazine.*)

Cable Size (Inches)	Maximum Frequency (MHz)	Velocity (%)	Peak Power (kW)	Average Power 1 MHz (kW)	Average Power 100 MHz (kW)	Attenuation* 1 MHz (dB)	Attenuation* 100 MHz (dB)
1–5/8	2.7	92.1	145	145	14.4	0.020	0.207
3	1.64	93.3	320	320	37	0.013	0.14
4	1.22	92	490	490	56	0.0010	0.113
5	0.96	93.1	765	765	73	0.007	0.079

*Attenuation specified in dB/100 ft.

The critical frequency for a chosen line length should not fall closer than 2 MHz to the pass band of the operating frequency.

Attenuation is related to the construction of the cable itself and varies with frequency, product dimension, and dielectric constant. Larger diameter cable exhibits lower attenuation than smaller diameter cable of similar construction when operated at the same frequency. It follows, therefore, that larger-diameter cables should be used for long runs.

Air-dielectric coaxial cable exhibits less attenuation than comparable sized foam-dielectric cable. The attenuation characteristic of a given cable is also affected by standing waves present on the line resulting from an impedance mismatch. Table 10-1 shows a representative sampling of semi-flexible coaxial cable specifications for a variety of line sizes.

10.2.3 Coaxial Cable Ratings

Selection of a type and size of transmission line is determined by a number of parameters, including: power-handling capability, attenuation and phase stability.

Power Rating. Both the peak and average power ratings are required to fully describe the capabilities of a transmission line. In most applications, the peak power rating limits the low frequency or pulse energy, while the average power rating limits high-frequency applications. Figure 10-2 charts the average power handling capabilities of a group of semi-flexible transmission lines.

The peak power rating is limited by the voltage breakdown potential between the inner and outer conductors of the line. The breakdown point is independent of frequency. It varies, however, with the

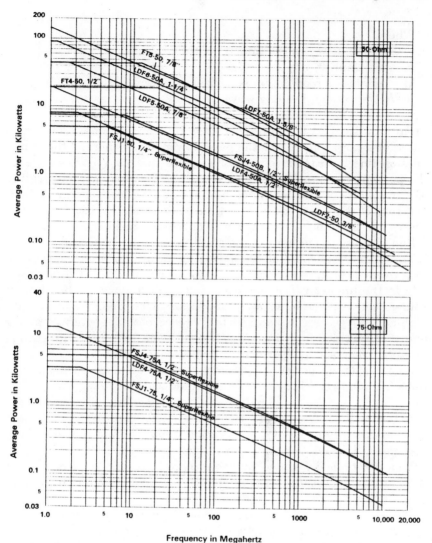

Figure 10-2 Average power ratings data for a variety of coaxial transmission lines. (Courtesy of Andrew Corp.)

line pressure (for an air-dielectric cable) and the type of pressurizing gas.

The peak power rating for a given system must satisfy the following equation:

$$P_{pk} > P_t (1 + M)^2 \text{ VSWR}$$

Where:

P_{pk} = peak power rating (in kW) of the cable
P_t = transmitter output power
M = amplitude modulation percentage (expressed decimally)
VSWR = voltage standing wave ratio

From this equation it can be seen that 100 percent amplitude modulation of a carrier increases the peak power requirement of the transmission line by a factor of four. Note also that as VSWR increases, the peak power rating must also increase. The average power rating of a transmission line is limited by the safe long-term operating temperature of the inner conductor and the dielectric. Excessive temperatures on the inner conductor will cause the dielectric material to soften, leading to mechanical instability inside the line.

The primary purpose of pressurization of an air-dielectric cable is to prevent the ingress of moisture. Moisture, if allowed to accumulate in the line, can increase attenuation and reduce the breakdown voltage between the inner and outer conductors. Pressurization with high-density gases can be used to increase both the average power and the peak power ratings of a transmission line. For a given line pressure, increased power rating is more significant for peak power than for average power. High-density gases used for such applications include Freon 116 and sulfur hexafluoride.

Attenuation. The attenuation characteristics of a transmission line vary as a function of the operating frequency and the size of the line itself. The relationships are shown in Figure 10-3. Attenuation values are typically stated by cable manufacturers for operation at 20°C. As temperature rises, attenuation rises. Attenuation also increases as a function of VSWR.

The *efficiency* of a transmission line dictates the amount of power output by the transmitter actually reaches the antenna. Efficiency is determined by the length of the line and the attenuation per unit length.

The attenuation of a coaxial transmission line is defined by the equation:

$$\alpha = 10 \times \log (P_1/P_2)$$

Where:

α = attenuation in dB/100 m
P_1 = input power into a 100 m line terminated with the nominal value of its characteristic impedance
P_2 = power measured at the end of the line

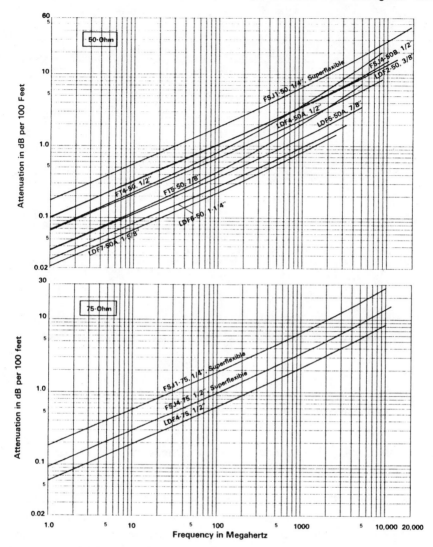

Figure 10-3 Attenuation characteristics for a selection of coaxial cables. (Courtesy of Andrew Corp.)

Stated in terms of efficiency.

Efficiency (%) = 100 (P_o/P_i)

Where:
P_i = power delivered to the input of the transmission line
P_o = power delivered to the antenna

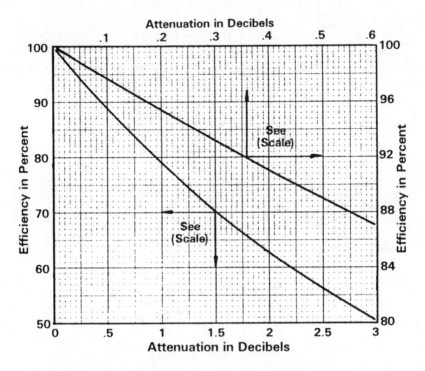

Figure 10-4 Conversion chart showing the relationship between decibel loss and efficiency of a transmission line. (Courtesy of Andrew Corp.)

The relationship between efficiency and loss in decibels (*insertion loss*) is shown in Figure 10-4.

Phase Stability. A coaxial cable expands as the temperature increases, causing the electrical length of the line to also increase. This factor results in phase changes that are a function of operating temperature. The phase relationship can be described by the equation:

$$\theta = 3.66 \times 10^{-7} \times P \times L \times T \times F$$

Where:

θ =   phase change in degrees
P =   phase temperature coefficient of the cable
L =   length of coax in feet
T =   temperature range (minimum-to-maximum operating temperature)
F =   frequency in MHz

Phase changes that are a function of temperature are important in systems that utilize multiple transmission lines, such as a directional array fed from a single phasor. To maintain proper operating parameters, the phase changes of the cables must be minimized. Specially-designed coaxial cables are available that offer low-phase-temperature characteristics. Two types of coax are available:

- *Phase stabilized.* Cables which have undergone extensive temperature cycling until such time as they exhibit their minimum phase temperature coefficient.
- *Phase compensated.* Cables in which changes in the electrical length have been minimized through adjustment of the mechanical properties of the dielectric and inner/outer conductors.

10.2.4 Mechanical Parameters

Corrugated copper cables are designed to withstand bending with no change in properties. Low-density foam- and air-dielectric cables generally have a minimum bending radius of ten times the cable diameter. So-called *super flexible* versions provide a much smaller bending radius. Rigid transmission line will not tolerate bending. Instead, transition elements (elbows) of various sizes are used. Individual sections of rigid line are secured by multiple bolts around the circumference of a coupling flange.

When a large cable must be used to meet attenuation requirements, short lengths of a smaller cable (jumpers or *pigtails*) may be used on either end for ease of installation in low power systems. The tradeoff is slightly higher attenuation and some additional cost.

The *tensile strength* of a cable is defined as the axial load that may be applied to the line with no more than 0.2 percent permanent deformation after the load is released. When divided by the weight per foot of cable, this gives an indication of the maximum length of cable that is self-supporting and therefore can be installed readily on a tower with a single hoisting grip. This consideration usually applies only to long runs of corrugated line; rigid line is installed one section at a time.

The *crush strength* of a cable is defined as the maximum force per linear inch that may be applied by a flat plate without causing more than a five percent deformation of the cable diameter. Crush strength is a good indicator of the ruggedness of a cable and its ability to withstand rough handling during installation.

Cable jacketing affords mechanical protection during installation and service. Semi-flexible cables are typically supplied with a jacket consisting of low-density polyethylene blended with three percent carbon black for protection from the sun's ultra-violet rays, which can degrade plastics given enough time. This approach has proven to be effective, yielding a life expectancy of more than 20 years. Rigid transmission line has no covering over the outer conductor.

For indoor applications, where fire-retardant properties are required, cables can be supplied with a fire-retardant jacket, usually listed by Underwriters Laboratories. Note that under the provisions of the National Electrical Code, outside plant cables such as standard black polyethylene-jacketed coaxial line may be run for as much as 50 ft. inside a building with no additional protection. The line also may be placed in conduit for longer runs.

Low-density foam cables are designed to prevent water from traveling along their length, should it enter through damage to the connector or the cable sheath. This is accomplished by mechanically locking the outer conductor to the foam dielectric by annular corrugations. Annular or ring corrugations, unlike helical or screw-thread type corrugations, provide a water block at each corrugation. Closed-cell polyethylene dielectric foam is bonded to the inner conductor, completing the moisture seal.

Installation Considerations. A coaxial cable is of no use whatsoever if it can't be reliably connected to other parts of the transmission system. Connector design, therefore, takes into account several key requirements. The connector interface must provide a weatherproof bond with the cable to prevent water from penetrating the connection. This is ensured by using O-ring seals. The cable-connector interface also must provide a good electrical bond that does not introduce a mismatch and increase VSWR. Good electrical contact between the connector and the cable also ensures that proper RF shielding is maintained.

Windloading is always a concern when a tower is designed. Overall cable diameter and the configuration in which it is hung effect the tower windload. This is especially important when several cable runs are required. If these runs are mounted side by side, they increase the windload. Windload reduction can be achieved by a cluster-mount arrangement in which the cables are mounted in a side-by-side circular configuration.

Proper installation of a cable system requires various accessories. The method used to hang a cable system determines both immediate and long-term performance. Two basic types of hangers are available for use with flexible cable.

- *Wrap-type hanger.* This hanger offers quick and easy installation, but there are some drawbacks to its use. Many such hangers are made of nylon and have limited mechanical durability. Others are made of malleable metal and are installed by twisting wires together. Such devices have a low resistance to cable slippage, and the clamping force may be inconsistent (determined by the installer).
- *Clamp hangers.* The clamp hanger is similar to a conduit-type hanger, but the halves are easier to spread apart. Clamp hangers require hardware for mounting to the tower bracket and for clamping the cable. This type of hanger provides dependable support and should be used wherever possible. The additional time required for installation is usually not significant.

A typical installation for semi-flexible cable is shown in Figure 10-5.

Installation of rigid line involves more complicated mechanical considerations (see Figure 10-6). Common hardware includes:

- *Bullets.* Connector elements used to electrically join the inner conductors of transmission line sections.
- *Elbows.* Pre-formed copper angle elements used to transition from one direction to another. Elbows are commonly available in 90° and 45° types.
- *Expansion joints.* A flexible-section inner conductor element used to accommodate vibration and differential expansion of the inner and outer conductors of a long line.
- *Line clamps.* Mounting hardware to secure the line to a tower leg or other supporting structure.
- *Spring hangers.* A flexible version of the line clamp, used to compensate for differential thermal expansion of the transmission line and the tower, or other supporting structure.

10.3 Waveguide

As the operating frequency of a system reaches into the UHF band, waveguide-based transmission line systems become practical. Waveguide is simplicity itself. There is no inner conductor; RF energy is *launched* into the structure and propagates to the load. Several types of waveguide are available, including: rectangular, square, circular and elliptical. Waveguide offers several advantages over coax. First, unlike coax, waveguide can carry more power as the operating frequency increases. Second, efficiency is significantly better with waveguide at higher frequencies.

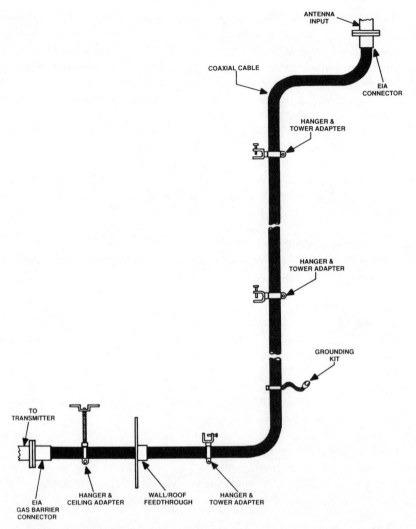

Figure 10-5 Typical installation hardware for a semi-flexible transmission line system. (Courtesy of Andrew Corp.)

Rectangular waveguide is commonly used in high-power transmission systems. Circular waveguide may also be used, especially for applications requiring a cylindrical member, such as a rotating joint for an antenna feed. The physical dimensions of the guide are selected to provide for propagation in the dominant (lowest-order) mode.

Waveguide is not without its drawbacks, however. Rectangular or square guide constitutes a large windload surface, which places sig-

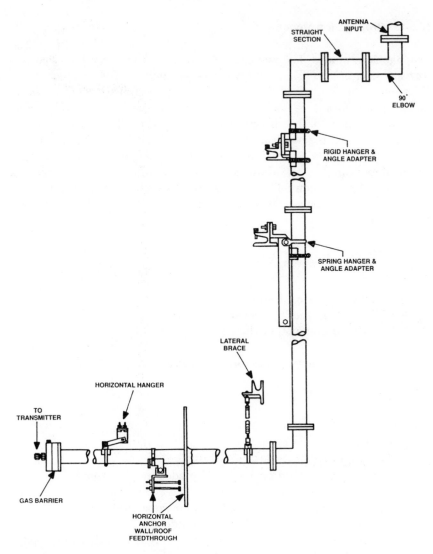

Figure 10-6 Typical installation hardware for a rigid transmission line system. (Courtesy of Andrew Corp.)

nificant structural demands on the tower. Because of the physical configuration of rectangular and square guide, pressurization is limited, depending on the type of waveguide used (0.5 psi is typical). Excessive pressure can deform the guide shape and result in increased VSWR. Wind may also cause deformation and ensuing VSWR problems. These considerations have led to the development of circular and elliptical waveguide.

10.3.1   Propagation Modes

Propagation modes for waveguide fall into two broad categories: *transverse-electric* (TE) and *transverse-magnetic* (TM) waves. With TE waves, the electric vector (*E vector*) is perpendicular to the direction of propagation. With TM waves, the magnetic vector (*H vector*) is perpendicular to the direction of propagation. These propagation modes take on integers (from 0 or 1 to infinity) that define field configurations. Only a limited number of these modes can be propagated, depending on the dimensions of the guide and the operating frequency. Energy cannot propagate in waveguide unless the operating frequency is above the *cutoff frequency*. The cutoff frequency for rectangular guide is:

$$F_c = c/2a$$

Where:
F_c = waveguide cutoff frequency
$c = 1.179 \times 10^{10}$ in/sec (the velocity of light)
a = the wide dimension of the guide

The cutoff frequency for circular waveguide is:

$$F_c = c/3.41a'$$

Where:
a' = the radius of the guide

There are four common propagation modes in waveguide:

- $TE_{0,1}$. The principle mode in rectangular waveguide.
- $TE_{1,0}$. Also used in rectangular waveguide.
- $TE_{1,1}$. The principle mode in circular waveguide. $TE_{1,1}$ develops a complex propagation pattern with electric vectors curving inside the guide. This mode exhibits the lowest cutoff frequency of all modes, which allows a smaller guide diameter for a specified operating frequency.
- $TM_{0,1}$. Has a slightly higher cutoff frequency than $TE_{1,1}$ for the same size guide. Developed as a result of discontinuities in the waveguide, such as flanges and transitions, $TM_{0,1}$ energy is not coupled out by either dominant or cross-polar transitions. The parasitic energy must be filtered out, or the waveguide diameter picked carefully to reduce the unwanted mode.

Waveguide will support dual-polarity transmission within a single run of line. A combining element (*dual-polarized transition*) is used at the beginning of the run, and a splitter (*polarized transition*) is used at the end of the line. Square waveguide has found numerous applications in such systems. The $TE_{1,0}$ and $TE_{0,1}$ modes are theoretically capable of propagation without cross coupling, at the same frequency, in lossless waveguide of square cross section. In practice, surface irregularities, manufacturing tolerances and wall losses give rise to $TE_{1,0}$ and $TE_{0,1}$ mode cross-conversion. Because this conversion occurs continually along the waveguide, long guide runs are usually avoided in dual-polarity systems.

Efficiency. Waveguide losses result from the following:

- Dissipation in the waveguide walls and the dielectric material filling the enclosed space
- Leakage through the walls and transition connections of the guide
- Localized power absorption and heating at the connection points

The operating power of waveguide may be increased through pressurization. Sulfur hexafluoride is commonly used as the pressurizing gas.

10.3.2 Ridged Waveguide

Rectangular waveguide may be ridged to provide a lower cutoff frequency, and thereby permit use over a wider frequency band. One- and two-ridged guides are used, as illustrated in Figure 10-7. Increased bandwidth comes at the expense of increased attenuation, relative to an equivalent section of rectangular guide.

10.3.3 Circular Waveguide

Circular waveguide offers several mechanical benefits over rectangular or square guide. The wind load of circular guide is 2/3 that of rectangular waveguide. It also presents more uniform windloading than rectangular waveguide, reducing tower structural requirements.

The same physical properties of circular waveguide that give it good power handling and low attenuation also result in electrical complexities. Circular waveguide has two potentially unwanted modes of propagation, the cross-polarized $TE_{1,1}$ and $TM_{0,1}$ modes. A circular waveguide, by definition, has no short or long dimension and, consequently, no method to prevent the development of *cross-polar* or *orthogonal* energy. Cross-polar energy is formed by small

A   B

Figure 10-7 Ridged waveguide: (a) single-ridged; (b) double-ridged.
(*Source:* Fink and Christiansen, *Electronics Engineers' Handbook,*
McGraw-Hill, New York, 1989.)

ellipticities in the waveguide. If the cross-polar energy is not trapped
out, the parasitic energy can recombine with the dominant mode energy.

Parasitic Energy. A hollow circular waveguide works as a high-Q
resonant cavity for some energy and as a transmission medium for
the rest. The parasitic energy present in the cavity formed by the
guide will appear as increased VSWR if not disposed of. The polarization in the guide meanders and rotates as it propagates from the
transmitter to the antenna. The end pieces of the guide, typically
circular-to-rectangular transitions, are polarization-sensitive. See the
top portion of Figure 10-8. If the polarization of the incidental energy
is not matched to the transition, energy will be reflected.

Several factors can result in this undesirable polarization. One
cause is out-of-round guides that result from non-standard manufacturing tolerances. In Figure 10-8, the solid lines depict the situation
at launching: perfectly circular guide with perpendicular polarization. The dashed lines show how certain ellipticities cause polarization rotation into unwanted states, while others have no effect. A 0.2
percent change in diameter can produce a -40 dB cross-polarization
component per wavelength. This is roughly 0.03 in. for 18 in. of
guide length.

Other sources of cross-polarization include twisted and bent
guides, out-of-roundness, offset flanges and transitions. Various
methods are used to dispose of this energy trapped in the cavity,
including absorbing loads placed at the ground and/or antenna level.

10.3.4   Doubly-Truncated Waveguide

The design of *doubly truncated waveguide* (DTW) is intended to overcome the problems that may result from parasitic energy in a circular waveguide. As shown in Figure 10-9, DTW consists of an almost

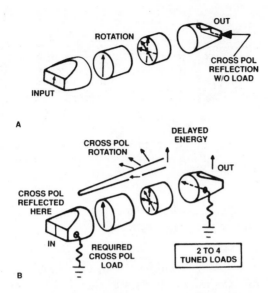

Figure 10-8 The effects of parasitic energy in circular waveguide: (a) trapped cross-polarization energy; (b) delayed transmission of the trapped energy. (*Source: Broadcast Engineering Magazine.*)

elliptical guide inside a circular shell. This guide does not support cross-polarization; tuners and absorbing loads are not required. The low windload of a hollow circular guide is maintained, except for the flange area.

Each length of waveguide is actually two separate pieces, a doubly truncated center section and a circular outer skin, joined at the flanges on each end. A large hole in the broadwall serves to pressurize the circular outer skin. Equal pressure inside the DTW and inside the circular skin ensures that the guide will not "breathe" or buckle as a result of rapid temperature changes. DTW exhibits about 3 percent higher windloading than circular waveguide (because of the transition section at the flange joints), and 32 percent lower loading than comparable rectangular waveguide.

10.3.5 Impedance Matching

The efficient flow of power from one type of transmission medium to another requires matching of the field patterns across the boundary to launch the wave into the second medium with a minimum of reflections. Coaxial line is typically matched into rectangular waveguide by extending the center conductor of the coax through the broadwall of the guide, parallel to the electric field lines, across the guide.

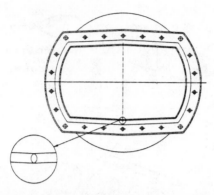

Figure 10-9 Physical construction of doubly-truncated waveguide. (Source: Broadcast Engineering Magazine.)

Alternatively, the center conductor can be formed into a loop and oriented to couple the magnetic field to the guide mode.

Waveguide Filters. A section of waveguide beyond cutoff constitutes a simple high-pass reflective filter. Loading elements in the form of posts or stubs may be employed to supply the reactances required for conventional lumped-constant-filter designs.

Absorption filters avoid the reflection of unwanted energy by incorporating lossy material in secondary guides that are coupled through so-called *leaky walls* (small sections of guide beyond cutoff in the passband). Such filters are typically used to suppress harmonic energy.

10.3.6 Installation Considerations

Waveguide system installation is both easier and more difficult than traditional transmission-line installation. There is no inner conductor to align, but alignment pins and more bolts are required per flange. Transition hardware to accommodate loads and coaxial-to-waveguide interfacing is also required. Figure 10-10 shows a typical dual-polarized microwave waveguide system.

Flange reflections can add up in phase at certain frequencies, resulting in high VSWR. The length of the guide must be chosen so that flange reflection build-up does not occur within the operating bandwidth.

Installing DTW is similar to installing circular or rectangular waveguide, as shown in Figure 10-11. With the exception of the rectangular "E" plane sweep at the bottom of the vertical run, the DTW

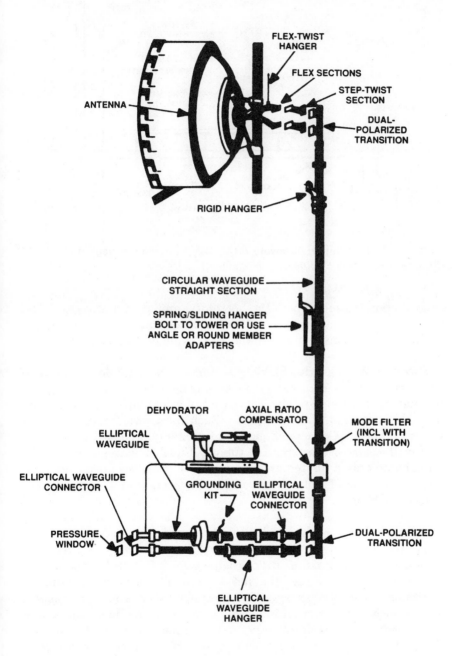

Figure 10-10 Typical installation of a dual-polarized microwave system.
(Courtesy of Andrew Corp.)

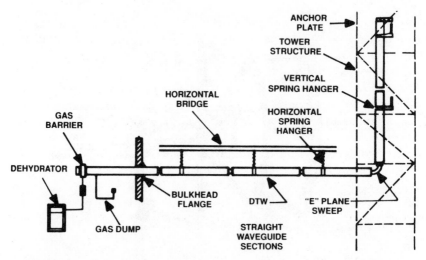

Figure 10-11 Installation hardware for doubly-truncated waveguide.
(*Source: Broadcast Engineering Magazine.*)

cross-section is used in both the horizontal and the vertical runs. The transition from rectangular to DTW on either end of the elbow occurs in the flanges. Constant-force hangers provide lateral support.

Flexible Waveguide. Flexible sections of waveguide are used to join rigid sections or components that cannot be otherwise aligned. Flexible sections also permit controlled physical movement resulting from thermal expansion of the line. Such hardware is available in a variety of forms. Corrugated guide is commonly produced by shaping thin-wall seamless rectangular tubing. Flexible waveguide can accommodate only a limited amount of mechanical movement. Depending on the type of link, the manufacturer may specify a maximum number of bends.

Tuning. Circular waveguide must be tuned. This process is a two-step procedure. First, the cross-polar $TE_{1,1}$ component is reduced, primarily through *axial ratio compensators* or *mode optimizers*. These devices counteract the net system ellipticity and indirectly minimize cross-polar energy. The cross-polar filters may also be rotated to achieve maximum isolation between the dominant and cross-polar modes. Cross-polar energy manifests itself as a net signal rotation at the end of the waveguide run. A perfect system would have a net rotation of zero. In the second step, tuning slugs at both the top and bottom of the waveguide run are adjusted to reduce the overall system VSWR. Tuning waveguide can be a complicated and time-consuming procedure; but normally, once set, tuning does not drift and must be redone only if major component changes are made.

Waveguide Hardware. Increased use of waveguide has led to the development of waveguide-based hardware for all elements, from the output of the transmitter to the antenna. Waveguide-based filters, elbows, directional couplers, switches, combiners and diplexers are currently available. Such hardware permits waveguide to be used from the output of the power-generating device to the antenna. Waveguide-based antennas are also available.

The RF performance of a waveguide component is usually better than the same item in coax. This is especially true in the case of diplexers and filters. Waveguide-based hardware provides lower attenuation and greater power-handing capability for a given physical size.

Bibliography

1. Perelman, R. and T. Sullivan, "Selecting Flexible Coaxial Cable," *Broadcast Engineering Magazine*, Overland Park, KS, May 1988.
2. Krohe, Gary L., "Using Circular Waveguide," *Broadcast Engineering*, Overland Park, KS, May 1986.
3. Ben-Dov, O. and C. Plummer, "Doubly Truncated Waveguide," *Broadcast Engineering*, Overland Park, KS, January 1989.
4. Andrew Corporation Bulletin 1063H, "Broadcast Transmission Line Systems," Orland Park, IL, 1982.
5. "Rigid Coaxial Transmission Lines," Cablewave Systems Catalog 700, North Haven, CT, 1989.
6. Cablewave Systems Technical Bulletin 21A, "The Broadcaster's Guide to Transmission Line Systems," North Haven, CT, 1976.
7. Fink, D. and D. Christiansen, *Electronics Engineer's Handbook*, Third Edition, McGraw-Hill, New York, 1989.
8. Vaughan, Thomas J., "RF Components for High Power and Super Power UHF-TV Transmitting Systems," *Proceedings of the International Broadcasting Convention*, Brighton, England, 1988.
9. "Circular Waveguide: System Planning, Installation and Tuning," Andrew Technical Bulletin 1061H, Orland Park, IL, 1980.
10. Jordan, Edward C., *Reference Data for Engineers: Radio, Electronics, Computer and Communications*, Seventh Edition, Howard W. Sams Company, Indianapolis, 1985.

11

RF Combiner Systems

11.1 Introduction[1]

There aren't enough skyscrapers judiciously placed to accommodate all the commercial and broadcast stations in need of a high perch for their antennas. And when high building sites are available, they are usually expensive. More often than not, a station must vie with other services for adequate high ground to erect an appropriate tower and transmitter building. Such restrictions have led to the widespread use of *community antenna sites*, where several stations are *multiplexed* (combined) into a single wideband radiator.

Use of a community site can result in significant cost savings for the participants. As a rule of thumb, a three-station partnership can bring each station's expenses down to a little less than it would cost for one station alone to build a transmitting facility without a multiplexer. For example, six class C (100 kW ERP) FM broadcasting stations using the same tower and multiplexer can reduce the start-up costs of a $1-million tower project to about $166,000 each. Each station's share of the $360,000 multiplexer would be $60,000, a small price to pay for more than $800,000 saved on the shared tower.

---

1 Portions of this chapter were adapted from: (1) DeComier, Bill, "Inside FM Multiplexer Systems," *Broadcast Engineering* Magazine, May 1988. (2) Heymans, Dennis, "Hot Switches and Combiners," *Broadcast Engineering Magazine*, December 1987.

11.1.1   Planning a Combiner System

Designing a multiplexer system requires specification of the following basic parameters:

- Power level that will be going through the multiplexer.
- The number of stations involved.
- Expansion capability of the multiplexer.
- Electrical performance requirements.
- Physical space available for the multiplexer. (The six-station FM system mentioned previously would require floor space of about 20 × 55 ft., or 1,100 sq. ft.).
- Number of output ports.
- Type of antenna feed.
- Status monitoring system.
- Failure mode override capability.

Meeting the last objective usually requires installation of a patch panel. In most high power multiplexer designs, routing around a failed module requires the ability to reroute the input power, not only at the input to the module, but around the output port as well. Because the module is symmetrical in many respects, it will pass power in the forward *and reverse* direction. The module must be taken out of the link in order to fully bypass it. To accomplish all of this requires a patch panel on the output side of the multiplexer, as well as the input side. Panels for this type of installation may be small and inexpensive for a low power system, or large and very expensive for a high power system. The six-station FM combiner would require a patch panel made of six or nine in. coax. Patch panels are the simplest way to bypass a module, but they are also the most expensive way. A less expensive approach involves use of an extra long *U-link* to bridge between the appropriate ports.

The type of monitoring system chosen for the facility is another important part of the planning process. Monitoring forward and reflected power at the input of each module can identify module failure. Monitoring forward and reflected power at the output of the combined system can identify antenna problems.

11.2 Combiner Types

The FM multiplexer discussed in the previous section represents RF combining on a grand scale. Most combiner systems are not so large and impressive. Combiners can be found in a wide variety of equip-

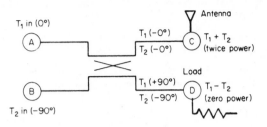

Figure 11-1 Operating principles of a 2-port hybrid combiner. This circuit is used to add two identical signals at inputs A and B. (Courtesy Micro Communications, Inc. in K. Blair Benson, *Television Engineering Handbook*, McGraw-Hill, New York, 1986.)

ment; most are hidden within the transmitter cabinet and treated as just another component in the RF chain.

In this chapter, the names *multiplexer* and *combiner* are used interchangeably to describe systems that mix signals in order to provide a common output.

11.2.1   2-Port Hybrid Combiner

The 2-port hybrid is the simplest combiner. Two signals are added to feed a common load. The hybrid presents a constant impedance to match each source. One RF unit must be phase-delayed by 90° to cancel out the inherent quadrature splitting by the combiner. Figure 11-1 shows a simplified 2-port hybrid combiner.

The combiner accepts one RF source and splits it equally into two parts, one arriving at output port C with 0° phase (no phase delay; it is the *reference phase*), and the other delayed by 90° at port D. A second RF source connected to input port B, but with a phase delay of 90°, will also split in two, but the signal arriving at port C will now be in-phase with source 1 and the signal arriving at port D will cancel, as shown in the figure.

Output port C, the summing point of the hybrid, is connected to the load. Output port D is connected to a resistive load to absorb any residual power resulting from slight differences in amplitude and/or phase between the two input sources. If one of the RF inputs should fail, half of the remaining transmitter output would be absorbed by the resistive load at port D.

The 2-port hybrid works only when the two signals being mixed are identical in frequency and amplitude, and when their relative phase is 90°. Operation of the hybrid can best be described by a *scat-*

Table 11-1 Single 90° hybrid system operating modes. (*Source: Broadcast Engineering Magazine.*)

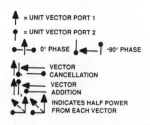

MODE	INPUT		SCHEMATIC	OUTPUT	
	1	2		3	4
1	$P_1 / 0°$	$P_2 \angle -90°$		0	$P_1 + P_2$
2	$P_1 / 0°$	$P_2 / 90°$		$P_1 + P_2$	0
3	$P_1 / 0°$	$P_2 / 0°$		$P_{1/2} + P_{2/2}$	$P_{1/2} + P_{2/2}$
4	$P_1 / 0°$	$P_2 = 0$		$P_{1/2}$	$P_{1/2}$
5	$P_1 = 0$	$P_2 / 0°$		$P_{2/2}$	$P_{2/2}$

↑ = UNIT VECTOR PORT 1

↑ = UNIT VECTOR PORT 2

0° PHASE   -90° PHASE

VECTOR CANCELLATION

VECTOR ADDITION

INDICATES HALF POWER FROM EACH VECTOR

tering matrix in which vectors are used to show how the device operates (see Table 11-1). In a 3-dB hybrid, two signals are fed to the inputs. An input signal at port 1 with 0° phase will arrive in phase at port 3, and at port 4 with a 90° lag (-90°) referenced to port 1. If the signal at port 2 already contains a 90° lag (-90° referenced to port 1), both input signals will combine in phase at port 4. The signal from port 2 also experiences another 90° change in the hybrid as it reaches port 3. Therefore, the signals from ports 1 and 2 cancel each other at port 3. If the signal arriving at port 2 leads by 90°, (mode 1 in the table), the combined power from ports 1 and 2 appears at port 4. If the two input signals are matched in phase (mode 4), the output

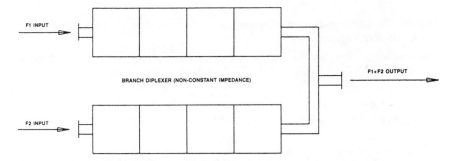

Figure 11-2 Non-constant impedance branch diplexer. In this configuration, two banks of filters feed into a coaxial tee. (*Source: Broadcast Engineering Magazine.*)

ports (3 and 4) contain one-half of the power from each of the inputs. If one of the inputs is removed, which would occur in a transmitter failure, only one hybrid input receives power (mode 5). Each output port would then receive 1/2 the input power of the remaining transmitter, as shown.

11.2.2 Non-Constant Impedance Diplexer

The *branch diplexer* is the typical configuration for a diplexer that does not exhibit constant impedance inputs. As shown in Figure 11-2, the branch diplexer consists of two banks of filters each feeding into a coaxial tee. The electrical length between each filter output and the centerline of the tee is frequency sensitive, but this fact is more of a tuning nuisance than a genuine user concern.

For this type of diplexer, electrical parameters are a function of filter characteristics. The VSWR, insertion loss, group delay, and rejection/isolation will be the same for the overall system as they are for the individual banks of cavities. The major limitation of this type of combiner is the degree of isolation that can be obtained for closely spaced channels.

Performance Specifications. The typical performance of a non-constant impedance diplexer operating at 100 MHz with two FM channels spaced 2.0 MHz apart is as follows:

- VSWR < 1.05:1 at the operating frequency (f_o) ($\pm$)200 kHz
- Insertion loss < 0.25 dB at f_o, < 0.30 dB at f_o $\pm$200 kHz
- Group delay < $\pm$25 ns at f_o $\pm$150 kHz
- Isolation > 50 dB for frequencies > $\pm$2 MHz from carrier

11.2.3   Constant Impedance Diplexer

The constant impedance diplexer employs 3-dB hybrids and filters with a terminating load on the isolated port. The filters in this type of combiner can be either notch-type (Section 11.2.4) or a bandpass-type (Section 11.2.5). The performance characteristics will be noticeably different for each design.

11.2.4   Band-Stop Diplexer

The band-stop (notch) constant impedance diplexer is configured as shown in Figure 11-3. For this design, the notch filters must have a high Q response to keep insertion loss low in the pass-band skirts. The high Q characteristic results in a sharp notch, typically providing 35-40 dB of notch depth at a carrier fequency of 100 MHz, falling to 14 dB at ±50 kHz and 8 dB at ±100kHz.

Because this bandwidth is inadequate for many applications, at least two cavities are located in each leg of the diplexer. They are stagger-tuned, one high and one low. With one cavity tuned 25 kHz high and the other 25 kHz low, a reject response of 50 dB is achieved at carrier, 65 dB at ±25 kHz and 20 dB at ±133 kHz (operating at a reference frequency of 100 MHz). With this dual cavity reject response in each leg of the band-stop diplexer system, the following analysis explains the key performance specifications.

Narrowband Input Performance. If frequency f1 is fed into the top left port of Figure 11-3, it will be split equally into the upper and lower leg of the diplexer. Each of these signals will reach the filters in their respective leg and be rejected/reflected back toward the input hybrid, recombine, and emerge through the lower left port, also known as the *wideband output.*

The VSWR looking into the f1 input is approximately 1.05:1 at all frequencies in the band. Within the bandwidth of the reject skirts, the observed VSWR is equal to the termination of the wideband output. Outside of ±250 kHz the signals will pass by the cavities, enter the right-most hybrid, recombine, and emerge into the dummy load. Consequently, the out-of-band VSWR is — in fact — the VSWR of the load.

The insertion loss from the f1 input to the wideband output is on the order of 0.1 dB at carrier ±125 kHz. This insertion loss depends on perfect reflection from the cavities. As the rejection diminishes on the skirts at ±150 kHz, the insertion loss from f1 to the wideband output rises.

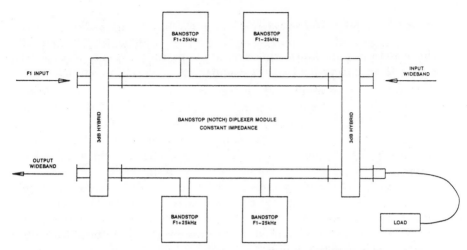

Figure 11-3 Band-stop (notch) constant impedance diplexer module. This design incorporates two 3 dB hybrids and filters, with a terminating load on the isolated port. (*Source: Broadcast Engineering Magazine.*)

The limitation in reject bandwidth of the cavities causes the insertion loss to rise at the edges. The isolation from f1 to the wideband input consists of a combination of the reject value of the cavities plus the isolation of the right-most hybrid.

A signal entering at f1 splits and proceeds in equal halves rightward through both the upper and lower legs of the diplexer. It is rejected by the filters 50 dB at carrier and 20 dB at ±130 kHz. Any residual signal that gets by the cavities reaches the right-most hybrid. There it recombines and emerges from the *load port*. The hybrid exhibits 35 dB isolation from the load port to the wideband input port. This hybrid isolation must be added to the filter rejection to obtain the total isolation from the f1 input to the wideband input.

Wideband Input Performance. If a signal is fed into the wideband input, the energy will split equally and proceed leftward along the upper and lower legs of the diplexer. Normally, f1 is not fed into the wideband input. If it was, f1 would be rejected by the notch filters and recombine into the load. All other signals, sufficiently removed from f1, will pass by the cavities with minimal insertion loss and recombine into the wideband output.

The VSWR looking into the wideband input is equal to the VSWR at the output for frequencies other than f1. If f1 were fed into the wideband input, the VSWR would be equal to the VSWR of the load.

Isolation from the wideband input to the f1 input is simply the isolation available in the left-most 3-dB hybrid, which is nominally 35 dB. This is usually inadequate for high power applications. To increase the isolation from the wideband input to the f1 input, it is necessary to use additional cavities between the f1 transmitter and the f1 input that will reject all frequencies fed into the wideband input by at least 20 dB. The 20 dB isolation of these cavities and the 35 dB of the hybrid results in a total isolation of 55 dB. This is usually adequate (with typical transmitter turn-around loss) to keep intermodulation products below the FCC limits.

Unfortunately, adding these cavities to the input line also cancels the constant impedance input. Although the notch diplexer shown in the figure is truly a constant impedance type of the diplexer, constant impedance is presented to the two inputs by virtue of using the hybrids at respective inputs.

The hybrids essentially cause the diplexer to act as an absorptive filter to all out-of-band signals. The out-of-band signals generated by the transmitter are absorbed by the load rather than reflected to the transmitter.

If a filter is added to the input to supplement isolation, the filter will reflect some out-of-band signals back at the transmitter. The transmitter will then be seeing the pass-band impedance of this supplemental filter rather than the constant impedance of the notch diplexer. This is a serious deficiency for applications that require a true constant impedance input.

11.2.5   Bandpass Constant Impedance Diplexer

The bandpass constant impedance diplexer is shown in Figure 11-4. This system takes all of the best features of diplexers and combines them into one unit. It also provides a constant impedance input that need not be supplemented with input cavities. Such cavities can rob the constant impedance diplexer of its constant impedance input.

The bandpass filters exhibit good bandwidth, providing (at a reference frequency of 100 MHz) 1.05:1 VSWR across ±200 kHz, insertion loss of 0.28 dB at carrier and rising to 0.30 dB at ±200 kHz. Rejection is 25 dB at ±800 kHz. This, when supplemented by isolation of the hybrids, will provide ample transmitter-to-transmitter isolation. Group delay is exceptional at ±25 ns for ±150 kHz. In fact, this system provides performance specifications similar to a branch-style bandpass system. It has the additional capability of providing port-to-port isolation of more than 50 dB between channels separated by only 800 kHz, as well as a true constant impedance input.

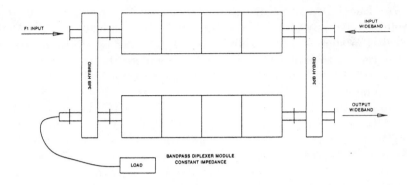

Figure 11-4 Bandpass constant impedance diplexer. This design takes the best features of other diplexers and combines them into one module. (*Source: Broadcast Engineering Magazine.*)

Narrowband Input Performance. The hybrids shown in the figure work in a manner identical to those described previously. However, the bandpass filters cause the system to exhibit performance specifications that exceed the band-stop system in every way. Consider a signal entering at the f1 input.

Within the pass-bands of the filters, which are also tuned to f1, the VSWR will be 1.05:1 at carrier, and <1.10:1 at f1 ±150 kHz. Because of the characteristic of the left-most hybrid, the VSWR is — in fact — a measure of the similarity of response of the top and bottom band of filters. The insertion loss looking from the f1 input to the wideband output will be similar to the insertion loss of the top and bottom filters individually. It is approximately 0.28 dB at carrier and 0.30 dB at ±200 kHz.

Both the insertion loss and group delay can be determined by the design bandwidth of the filters. Increasing bandwidth causes the insertion loss and the group delay deviation to decrease. Unfortunately, as the bandwidth increases with a given number of cavities, isolation suffers for closely-spaced channels because the reject skirt of the filter decreases with increasing bandwidth.

Isolation of f1 to the wideband input is determined as follows: a signal enters at the f1 input, splits equally into the upper and lower banks of filters, passes with minimal loss through the filters, and recombines into the wideband output of the right-most hybrid. Both the load and the wideband input ports are isolated by their respective hybrids to 30 dB below the f1 input level. Thus, isolation of the f1 input to the wideband input is 30 dB, which is inadequate. Fortunately, it will be supplemented by the reject skirt of the next module.

Wideband Input Performance. A signal fed into the wideband input could be any frequency at least 800 kHz removed from f1. As it enters, the signal will be split into equal halves by the hybrid and then proceed to the left until the two components reach the reject skirts of the filters. The filters will shunt all frequencies 800 kHz or more removed from f1. If the shunt energy is in phase for the given frequency when the signal is reflected back to the right hybrid, it will recombine into the wideband output. The VSWR under these conditions will be equal to the termination at the wideband output, typically <1.10:1. If the reject skirts of the filter are sufficient (>20 dB), the insertion loss from the wideband input to the wideband output will be 0.03 dB.

The isolation from the wideband input to f1 can be determined as follows: a signal enters at the wideband input, splits equally into upper and lower filters, and is rejected by the filters by at least 25 dB. Any residual signal that passes through the filters in spite of the 25 dB rejection will still be in the proper phase to recombine into the load, producing 30 dB isolation to the f1 input port. Thus, the isolation of the wideband input to the f1 input is the sum of the 25 dB from the filters and 30 dB from the left hybrid, for a total of 55 dB.

Isolation of f1 to f2. Extending the use of the diplexer module into a multiplexer application supplements the deficient isolation described previously (*narrowband input performance*), while maintaining the constant impedance input. In a multiplexer system the wideband input of one module is connected to the wideband output of the next module. (See Figure 11-5.)

It has already been stated that the isolation from the f1 input to the wideband input is deficient; however, additional isolation is provided by the isolation of the wideband output to the f2 input of the next module. Consider that f1 has already experienced 30 dB isolation to the wideband input of the same module. When this signal continues to the next module through the wideband output of module 2, it will be split into equal halves and proceed to the left of module 2 until it reaches the reject skirts of the filters in module 2.

These filters are tuned to f2 and reject f1 by at least 25 dB. The combined total isolation of f1 to f2 is the sum of the 30 dB of the right hybrid in module 1, plus the 25 dB of the reject skirts of module 2, for a total of 55 dB.

Intermodulation Products. The isolation just described is equal in magnitude to that of a band-stop module, but provides further protection against the generation of intermodulation products. The

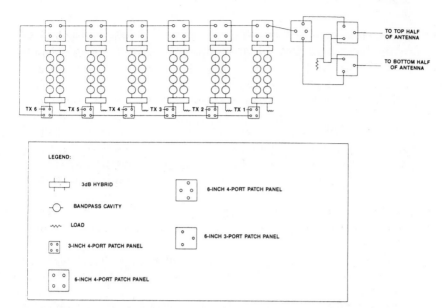

Figure 11-5 Schematic diagram of a 6-module bandpass multiplexer. This configuration accommodates a split antenna design and incorporates patch panels for bypass purposes. (*Source: Broadcast Engineering Magazine.*)

most troublesome intermod products usually occur when an incoming (secondary) signal mixes with the second harmonic of a primary transmitter.

When the primary transmitter is operating on frequency A, the intermod will occur at the frequency $(2 \times A) - B$. This formula invariably places the intermod from the primary transmitter symmetrically about the operating frequency. By an interesting coincidence, the bandpass filters in the bandpass module also provide symmetrical reject response on both sides of the primary operating frequency.

Notice that the incoming signal is attenuated by 30 dB in the respective hybrid and by 25 dB in the filter, for a total of 55 dB. If an intermod is still generated in spite of this isolation, it will emerge on the other skirt of the filter attenuated by 25 dB. In the bandpass system, the incoming signal is attenuated by 55 dB and the resulting outgoing spur by 25 dB. The total 80 dB suppression is equal to the 80 dB intermod requirement typically specified by the FCC.

Interestingly, the entire 80 dB of attenuation is supplied by the diplexer regardless of the so-called *turn-around loss* of the transmitter. The tendency toward wideband final stage amplifiers in transmitters requires constant impedance inputs. The transmitters also require increased isolation because they offer limited turn-around loss.

Group Delay. Group delay in the bandpass multiplexer module is equal to the sum of the narrowband and wideband input group delay of all modules between the input and the antenna. The narrowband input group delay is a U-shaped response, with minimum at center and rising to a maximum on both sides at the frequency where the reject rises to 3 dB. Group delay then decreases rapidly at first, and then very slowly.

If the bandwidth of the pass-band is made so that the group delay is ±25 ns over ±150 kHz (in a system operating near 100 MHz), the 3 dB points will be at ±400 kHz, and the out-of-band group delay will fall rapidly at ±800 kHz and possibly ±1.0 MHz. If there are no frequencies offset ±800 kHz or ±1.0 MHz up-stream in other modules, this poses no problem.

If there are modules upstream tuned to 800 kHz or 1.0 MHz on either side, then the group delay (when viewed at the up-stream module) will consist of its own narrowband input group delay plus the rapidly falling group delay of the wideband input of the closely-spaced down-stream module. Under these circumstances, if good group delay is desired, it is possible to utilize a *group delay compensation* module.

A group delay compensation module consists of a hybrid and two cavities used as notch cavities. It typically provides a group delay response that is inverted compared to a narrowband input group delay. Because group delay is additive, the inverted response subtracts from the standard response, effectively reducing the group delay deviation to within ±25 ns for ±150 kHz.

It should be noted that the improvement in group delay is obtained at the cost of insertion loss. In large systems (eight to ten modules), the insertion loss can be high because of the cumulative total of all wideband losses. Under those conditions, it may be more prudent to accept higher group delay and retain minimum insertion loss.

Table 11-2 lists typical electrical performance specifications for the branch, bandstop, and bandpass diplexer modules as applied to FM broadcast combining systems.

11.2.6   Microwave Combiners

Hybrid combiners are typically used in microwave amplifiers to combine the output energy of individual solid state modules to provide

Table 11-2 Performance characteristics of three multiplexer configurations, each using 4 cavity designs. (*Source: Broadcast Engineering Magazine.*)

Characteristic	4-Cavity Bandpass Branch	4-Cavity Bandstop Constant Impedance	8-Cavity Bandpass Constant Impedance
VSWR Bandwidth	±200 kHz	±130k Hz w/dual cavities	±200 kHz
VSWR	1.05:1	1.05:1	1.05:1
Insertion Loss BW	±200 kHz	±130k Hz w/dual cavities	±200 kHz
Insertion loss	−0.25 dB	−0.15 dB/NB input[1] −0.25 dB/WB input[2]	−0.25 dB/NB input −0.03 dB/WB input
Isolation	>50 dB	>50 dB	>50 dB
Nearest adjacent frequency	1.8 MHz	0.8 MHz	0.8 MHz
Needs supplemental cavities?	No	Yes	No
Effective constant impedance input?	No	No	Yes
Group delay BW	±150 kHz	±150 kHz	±150 kHz
Group delay deviation	±25 ns	±100 ns	±25 ns
Group delay symmetry?	Yes	No	Yes
Group delay correctable?	Yes	No	Yes
Symmetrical intermod protection?	Yes	No	Yes

[1] narrowband input
[2] wideband input

the necessary power output from the transmitter. Quadrature hybrids effect a VSWR-canceling phenomenon that results in well-matched power amplifier inputs, and outputs that can be broadbanded with the proper selection of hybrid trees.

Several hybrid configurations are possible, including the *split-T, branch-line, magic-T,* and *backward-wave.* Key design parameters include coupling bandwidth, isolation, and fabrication ease. The equal-amplitude quadrature-phase, reverse-coupled TEM 1/4-wave hybrid is particularly attractive because of its bandwidth and amenability to

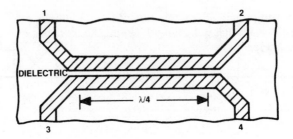

Figure 11-6 Reverse-coupled, 1/4-wave hybrid combiner.

various physical implementations. Such a device is illustrated in Figure 11-6.

11.3 Hot Switching Combiners

Switching RF is nothing new. Typically, the process involves coaxial switches, coupled with the necessary logic to ensure that the "switch" takes place with no RF energy on the contacts. This process usually takes the station off the air for a few seconds while the switch is completed. Through the use of hybrid combiners, however, it is possible to redirect RF signals without turning the carrier off. This process is referred to as *hot switching*. Figure 11-7 illustrates two of the most common switching functions (SPST and DPDT) available from hot switchers.

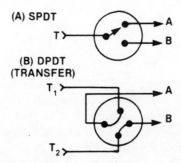

Figure 11-7 Common RF switching configurations. (*Source: Broadcast Engineering Magazine.*)

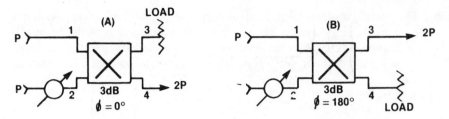

Figure 11-8 Hybrid switching configurations: (a) phase set so that the combined energy is delivered to port 4; (b) phase set so that the combined energy is delivered to port 3. *(Source: Broadcast Engineering Magazine.)*

11.3.1 Switch Types

The unique phase-related properties of an RF hybrid make it possible to use the device as a switch. The input signals to the hybrid in Figure 11-8 (a) are equally powered but differ in phase by 90°. This phase difference results in combined signals being routed to the output terminal at port 4. If the relative phase between the two input signals is changed by 180°, the summed output then appears on port 3, as shown in Figure 11-8 (b). The 3-dB hybrid combiner has just become a switch.

This configuration permits the switching of two transmitters to either of two antennas. Remember however, that the switch takes place when the phase difference between the two inputs is 90°. To perform the switch in a useful way requires adding a high-power phase shifter to one input leg of the hybrid. The addition of the phase shifter permits the full power to be combined and switched to either output.

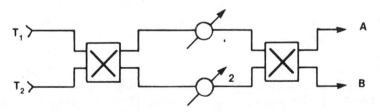

Figure 11-9 Additional switching and combining functions can be developed by adding a second hybrid and another phase shifter. *(Source: Broadcast Engineering Magazine.)*

This configuration of hybrid and phase shifter, however, will not permit switching a main or standby transmitter to a main or auxiliary antenna (DPDT function). To accomplish this additional function, a second hybrid and phase shifter must be added, as shown in Figure 11-9. This configuration can then perform the following switching functions:

- Transmitter 1 routed to output B
- Transmitter 2 routed to output A
- Transmitter 1 routed to output A
- Transmitter 2 routed to output B

The key element in developing such a switch is a reflectionless high-power phase shifter. In this application, the phase shifter allows the line between the hybrids to be electrically lengthened or shortened. The ability to adjust the relative phase of the two input signals to the second hybrid provides the needed control to switch the input signal between the two output ports (3 and 4).

If a continuous analog phase shifter is used, the transfer switch shown in the previous figure can also act as a hot switchless combiner where transmitters 1 and 2 can be combined and fed to either output A or B. The switching or combining functions are accomplished by changing the physical position of the phase shifter.

Note that it doesn't matter whether the phase shifter is in one or both legs of the system. It is the phase difference ($\theta 1 - \theta 2$) between the two input legs of the second hybrid that is important. With two-phase shifters, dual drives are required. However, the phase shifter need have only two positions. In a one-phase shifter design, only single drive is required but the phase shifter must have four positions.

Phase Relationships. To better understand the dual hybrid switching and combining process, examine Table 11-3. The table lists the various combinations of inputs, relative phase, and output configurations that are possible with the single phase shifter design.

Using vector analysis, note that when two input signals arrive in phase (mode 1) at ports 1 and 2 with the phase shifter set to 0°, the circuit acts like a crossover network with the power from input port 1 routed to output port 4. Power from input port 2 is routed to output port 3. If the phase shifter is set to 180°, the routing changes, with port 1 being routed to port 3 and port 2 being routed to port 4.

Mode 2 represents the case where one of the dual-input transmitters has failed. The output signal from the first hybrid arrives at the input to the second hybrid with a 90° phase difference. Because the

Table 11-3 Dual 90° hybrids and single phase-shifter combiner operating modes. (*Source: Broadcast Engineering Magazine.*)

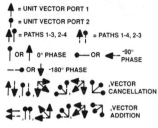

MODE	INPU. 1	INPU. 2	Ø	INPUT VECTOR	SCHEMATIC	OUTPUT VECTOR	OUTPUT 3	OUTPUT 4
1	P_1/ 0°	P_2/ 0°	0°				P_2	P_1
	P_1/ 0°	P_2/ 0°	180°				P_1	P_2
2	P_1/ 0°	P_2 = 0	0°				0	P_1
3	P_1 = 0	P_2/ 0°	0°				P_2	0
4	P_1/ 0°	P_2 = 0	180°				P_1	0
5	P_1 = 0	P_2/ 0°	180°				0	P_2
6	P_1/ 0°	P_2/ 0°	90°				0	$P_1 + P_2$
	P_1/ 0°	P_2/ 0°	- 90°				$P_1 + P_2$	0

second hybrid introduces a 90° phase shift, the vectors add at port 4 and cancel at port 3. This effectively switches the working transmitter connected to port 1 to output port 4, the antenna.

By introducing a 180° phase shift between the hybrids, as shown in modes 4 and 5, it is possible to reverse the circuit. This allows the outputs to be on the same side of the circuit as the inputs. This configuration might be useful in the event that transmitter 1 failed and all of the power from transmitter 2 had to be directed to a diplexer connected to output 4.

Normal operating configurations are shown in modes 6 and 7. When both transmitters are running, it is possible to have the combined power routed to either output port. The switching is accomplished by introducing a ±90° phase shift between the hybrids.

As shown by the table, it is possible to operate in all of the listed modes through the use of a single phase shifter. The phase shifter must provide four different phase positions. A similar analysis would show that a two-phase shifter design, with two positions for each shifter, is capable of providing the same operational modes.

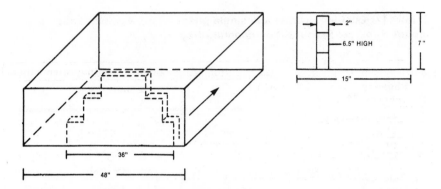

Figure 11-10 The dielectric vane switcher consists of a long dielectric sheet, mounted within a section of rectangular waveguide. (*Source: Broadcast Engineering Magazine.*)

11.3.2   Phase Shifters

The key to making hybrid switches work in the real world lies in the phase shifter. The dual 90° hybrid combiner just discussed requires a phase shifter capable of introducing a fixed phase offset of -90°, 0°, +90°, and +180°. This can be easily accomplished at low power levels through the use of a sliding short circuit (trombone type) line stretcher. However, when high frequency and high power signals are being used, the sliding short circuit is not an appropriate design choice. In a typical case, the phase shifter must be able to handle as much as 120 kW at UHF frequencies. Under these conditions, sliding short circuit designs are often unreliable. Therefore, three other methods have been developed: *variable dielectric vane, dielectric posts*, and *variable phase hybrid*.

Variable Dielectric Vane. The variable dielectric vane consists of a long dielectric sheet mounted in a section of rectangular waveguide (see Figure 11-10). The dielectric sheet is long enough to introduce a 270° phase shift when located in the center of the waveguide. As the dielectric sheet is moved toward the wall, into the lower field, the phase shift decreases. A single-sided phase shifter can easily provide the needed four positions. A two-stage quarter-wave transformer is used on each end of the sheet to maintain a proper match for any position over the desired operating band. The performance of a typical switchless combiner using the dielectric vane is shown in Table 11-4.

**Table 11-4 Performance of a dielectric vane phase-shifter.
(*Source: Broadcast Engineering Magazine.*)**

Single Input	Δ Degrees	VSWR	Input Attenuation		Output Attenuation	
			1 (dB)	2 (dB)	3 (dB)	4 (dB)
T1	180°	1.06	—	39	0.1	39
T1	0°	1.05	—	39	39	0.1
T2	180°	1.05	39	—	39	0.1
T2	0°	1.06	39	—	0.1	39
Dual Input						
T1 + T2	270°	1.06	—	—	0.1	36
T1 + T2	90°	1.06	—	—	36	0.1

Dielectric Posts. Dielectric posts, shown in Figure 11-11, operate on the same principle as the dielectric vane. The dielectric posts are positioned a quarter-wavelength apart from each other to cancel any mismatch, and maintain minimum VSWR.

Variable Phase Hybrid. The variable-phase hybrid, shown in Figure 11-12, relies on a 90° hybrid, similar to those used in a combiner. With a unit vector incident on port 1, the power is split by the 90° hybrid. The signal, at ports 3 and 4 are reflected by the short circuit. These reflected signals are out of phase at port 1 and in phase at port 2.

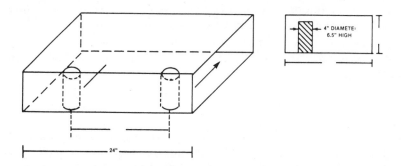

Figure 11-11 Dielectric posts are similar to dielectric vanes. The posts move within the waveguide, providing the necessary phase shift. (*Source: Broadcast Engineering Magazine.*)

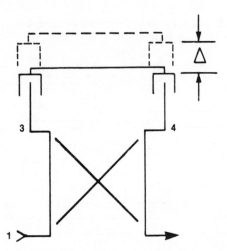

Figure 11-12 Variable-phase hybrid designs rely on a 90° hybrid and non-contacting-type short circuits. (*Source: Broadcast Engineering Magazine.*)

The relative phase of the hybrid can be changed by moving the short circuit. Mathematically, the system can be described by:

$$\theta = \frac{2\pi}{g} \times 2\Delta$$

Where:
Δ = the length of the shorting element

The variable-phase hybrid is linear with respect to position. Non-contacting choke-type short circuits, with high front-to-back ratios are typically used in the devices. The performance available from a typical high-power variable-phase switchless combiner is shown in Table 11-5.

11.4 High Power Isolators

The high power ferrite isolator offers the ability to stabilize imped-ance, isolate the transmitter, eliminate antenna reflections, and ab-sorb harmonic and intermodulation products. The isolator can also be used to switch between an antenna or load under full power, or combine two or more transmitters into a common antenna. Isolators are commonly used in microwave transmitters at low power to pro-tect the output stage from reflections. Until recently, however, the insertion loss of the ferrite made use of isolators impractical at high power levels (10 kW and above). Ferrite isolators are now available

Table 11-5 Performance of a variable-phase hybrid phase-shifter. (*Source: Broadcast Engineering Magazine.*)

Single Input	Δ Degrees	VSWR	Input Attenuation		Output Attenuation	
			1 (dB)	2 (dB)	3 (dB)	4 (dB)
T1	180°	1.06	—	36	0.1	52
T1	0°	1.04	—	36	50	0.1
T2	180°	1.06	36	—	52	0.1
T2	0°	1.07	36	—	0.1	50
Dual Input						
T1 + T2	270°	1.06	—	—	0.1	36
T1 + T2	90°	1.06	—	—	36	0.1

that can handle as much as 400 kW of forward power with less than 0.1 dB of forward power loss. Experimental models have been demonstrated that can handle 4 MW.

11.4.1 Theory of Operation

High power isolators are 3-port versions of a family of devices known as *circulators*. The circulator derives its name from the fact that a signal applied to one of the input ports can travel in only one direction, as shown in Figure 11-13. The input port is isolated from the output port. A signal entering port 1 appears only at port 2; it does not appear at port 3 unless reflected from port 2. An important benefit of this one-way power transfer is that the input VSWR at port 1 is dependent only on the VSWR of the load placed at port 3. In most applications, this load is a resistive (dummy) load that represents a perfect load to the transmitter.

The unidirectional property of the isolator results from the magnetization of the ferrite alloy inside the unit. By polarizing the magnetic field of the ferrite correctly, RF energy will travel through the element in only one direction (port 1 to 2, port 2 to 3, and port 3 to 1). By reversing the polarity of the magnetic field, RF flow in the opposite direction is possible. Recent developments in ferrite technology have resulted in high isolation with low insertion loss.

The ferrite is placed in the center of a Y-junction of three transmission lines, either waveguide or coax. Sections of the material are bonded together to form a thin cylinder perpendicular to the electric field. Even though the insertion loss is low, the resulting power dissipated in the cylinder can be as high as 2 percent of the forward power. Special provisions must be made for heat removal.

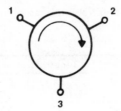

Figure 11-13 Basic characteristics of a circulator.

Values of ferrite loss on the order of 0.05 dB have been demonstrated. This equates to an efficiency of 98.9 percent. Additional losses from the transmission line and matching structure contribute slightly to loss. The overall loss is typically less than 0.1 dB, or 98 percent efficiency. The ferrite element is usually water-cooled in a closed-loop system that uses an external radiator.

11.4.2   Applications

The high power isolator permits a transmitter to operate with high performance and reliability despite a load that is less than optimum. The problems presented by ice formation on a transmitting antenna provide a convenient example. Ice buildup will detune an antenna, resulting in reflections back to the transmitter and high VSWR. If the VSWR is severe enough, transmitter power will have to be reduced to keep the system on the air. An isolator, however, permits continued operation with no degradation in signal quality. Power output is affected only to the extent of the reflected energy, which is dissipated in the resistive load.

A high power isolator can also be used to provide a stable impedance for devices that are sensitive to load variations, such as klystrons. This allows the device to be tuned for optimum performance regardless of the stability of the RF components located after the isolator. Figure 11-14 shows the output of a wideband (6 MHz) klystron operating into a resistive load, and into an antenna system. The power loss is the result of an impedance difference. The periodicity of the ripple shown in the trace is a function of the distance of the reflections from the source.

Hot Switch. The circulator can be made to perform a switching function if a short circuit is placed at the output port. Under this situation, all input power will be reflected back into the third port. The use of a non-contacting high power stub on port 2 permits redirecting the transmitter's output.

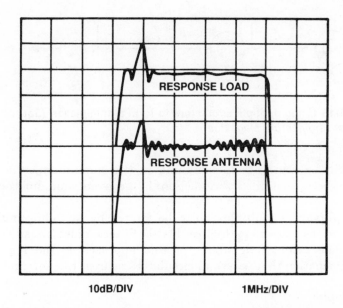

Figure 11-14 Output of a klystron operating into a resistive load, and into an antenna system. (Courtesy of Micro Communications, Inc.)

At odd quarter wave positions, the stub appears as a high imped-ance and has no effect on the output port. At even quarter-wave posi-tions, the stub appears as a short circuit. Switching between the an-tenna and a test load can, therefore, be accomplished by moving the short 1/4-wavelength.

Diplexer. An isolator may be configured to combine the aural and visual outputs of a television transmitter into a single output for the antenna. The approach is shown in Figure 11-15. A single notch cav-ity at the aural frequency is placed on the visual transmitter output (circulator input) and the aural signal is added (as shown). The aural

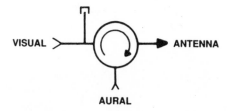

Figure 11-15 Use of a circulator as a diplexer in television applications. (Courtesy of Micro Communications, Inc.)

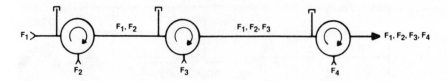

Figure 11-16 Using multiple circulators to form a multiplexer. (*Source: Broadcast Engineering Magazine.*)

signal will be routed to the antenna in the same manner as it is reflected (because of the hybrid action) in a conventional diplexer.

Multiplexer. A multiplexer can be formed by cascading multiple circulators, as shown in Figure 11-16. Filters must be added, as shown. The primary drawback of this approach is the increased power dissipation that occurs in circulators nearest the antenna.

Bibliography

1. Fink, D. and D. Christiansen, *Electronics Engineer's Handbook,* Third Edition, McGraw-Hill, New York, 1989.
2. Stenberg, James T., "Using Super Power Isolators in the Broadcast Plant," *Proceedings of the Broadcast Engineering Conference,* Denver, CO, 1988.
3. Vaughan, T. and E. Pivit, "High Power Isolator for UHF Television," *Proceedings of the NAB Engineering Conference,* Las Vegas, NV, 1989.

12

Antenna Systems

12.1 Introduction

Transmission is accomplished by the emission of coherent electromagnetic waves in free space from one or more radiating elements that are excited by RF currents. Although, by definition, the radiated energy is composed of mutually dependent magnetic and electric vector fields, it is conventional practice to measure and specify radiation characteristics in terms of the electric field only.

The purpose of an antenna is to radiate efficiently the power supplied to it by the transmitter. A simple antenna, consisting of a single vertical element over a ground plane can do this job quite well at low-to-medium frequencies. Antenna systems may also be required to concentrate the radiated power in a given direction and minimize radiation in the direction of other stations sharing the same or adjacent frequencies. To achieve such directionality may require a complicated antenna system that incorporates a number of individual elements or towers and matching networks.

As the operating frequency increases into VHF and above, the short wavelengths permit the design of specialized antennas that offer high directivity and gain.

12.1.1 Operating Characteristics

Wavelength is the distance traveled by one cycle of a radiated electric signal. The frequency of the signal is the number of cycles per second. It follows that the frequency is inversely proportional to the wavelength. Both wavelength and frequency are related to the speed of light. Conversion between the two parameters can be accomplished with the formula:

$$c = f \times w$$

Where:
c = speed of light
f = operating frequency
w = wavelength

The velocity of electric signals in air is essentially the same as that of light in free space (2.9983×10^{10} cm/s).

The *electrical length* of a radiating element is the most basic parameter of an antenna.

$$H° = \frac{Ht \times (Fo)}{2733}$$

Where:
H° = length of the radiating element in electrical degrees
Ht = length of the radiating element in feet
Fo = frequency of operation in kHz

Where the radiating element is measured in meters:

$$H° = \frac{Ht \times (Fo)}{833.23}$$

The *radiation resistance* of an antenna is defined by the equation:

$$R = \frac{P}{I^2}$$

Where:
R = radiation resistance
P = power delivered to the antenna
I = driving current at the antenna base

Antenna Bandwidth. Bandwidth is a general classification of the frequencies over which an antenna is effective. This parameter requires specification of acceptable tolerances relating to the uniformity of response over the intended operating band.

Strictly speaking, *antenna bandwidth* is the difference in frequency between two points at which the power output of the transmitter drops to one-half the midrange value. The points are called *half-power points*. A half-power point is equal to a VSWR of 5.83:1, or the point at which the voltage response drops to 0.7071 of the midrange value. In most communications systems, a VSWR of less than 1.2:1 within the occupied bandwidth of the radiated signal is preferable. Antenna bandwidth depends upon the radiating element impedance and the rate at which the reactance of the antenna changes with frequency.

Bandwidth and RF coupling go hand in hand regardless of the method used to excite the antenna. All elements between the transmitter output circuit and the antenna must be analyzed, first by themselves, and then as part of the overall system bandwidth. In any transmission system, the *composite bandwidth*, not just the bandwidths of individual components, is of primary concern.

Polarization. Polarization is the angle of the radiated electric field vector in the direction of maximum radiation. Antennas may be designed to provide horizontal, vertical, or circular polarization. Horizontal and vertical polarization is determined by the orientation of the radiating element with respect to Earth. If the plane of the radiated field is parallel to the ground, the signal is said to be horizontally polarized. If it is at right angles to the ground, it is said to be vertically polarized. When the receiving antenna is located in the same plane as the transmitting antenna, the received signal strength will be maximum.

Circular polarization (CP) of the transmitted signal results when equal electrical fields in the vertical and horizontal planes of radiation are out-of-phase by 90° and are rotating a full 360° in one wavelength of the operating frequency. The rotation can be clockwise or counterclockwise, depending on the antenna design. This continuously rotating field gives CP good signal penetration capabilities because it can be received efficiently by an antenna of any random orientation. Figure 12-1 illustrates the principles of circular polarization.

Antenna Beamwidth. Beamwidth in the plane of the antenna is the angular width of the directivity pattern where the power level of the received signal is down by 50 percent (3 dB) from the maximum signal in the desired direction of reception.

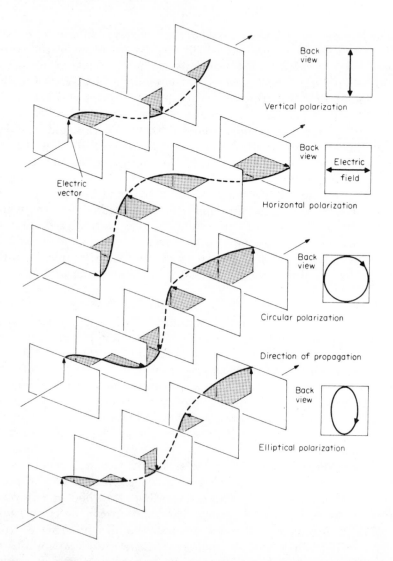

Figure 12-1 Polarization of the electric field of a transmitted wave.
(*Source:* K. Blair Benson, *Television Engineering Handbook,* McGraw-Hill, New York, 1986.)

Antenna Gain. *Directivity* and *gain* are measures of how well energy is concentrated in a given direction. Directivity is the ratio of power density in a given direction to the power density that would be produced if the energy were radiated isotropically. The reference can be linearly or circularly polarized. Directivity is usually given in *dBi* (decibels above isotropic).

Gain is the field intensity produced in a given direction by a fixed input power to the antenna, referenced to a dipole. Gain is frequently used as a figure of merit. Gain is closely related to directivity, which in turn is dependent upon the radiation pattern. High values of gain are usually obtained with a reduction in beamwidth.

An antenna is typically configured to exhibit "gain" by narrowing the beamwidth of the radiated signal to concentrate energy toward the intended coverage area. The actual amount of energy being radiated is the same with a unity gain antenna or a high gain antenna, but the useful energy (commonly referred to as the *effective radiated power*, or ERP) can be increased significantly.

Electrical *beam tilt* may also be designed into a high gain antenna. A conventional antenna typically radiates more than half of its energy above the horizon. This energy is lost for practical purposes in most applications. Electrical beam tilt, caused by delaying the RF current to the lower elements of a multi-element antenna, can be used to provide more useful power to the service area.

Pattern optimization is another method that may be used to maximize radiation to the intended service area. The characteristics of the transmitting antenna are sometimes greatly affected by the presence of the supporting tower, if side-mounted, or by nearby tall obstructions (such as another transmitting tower) if top-mounted. Antenna manufacturers use various methods to reduce pattern distortions. These generally involve changing the orientation of the radiators with respect to the tower and adding parasitic elements.

Space Regions. Insofar as the transmitting antenna is concerned, space is divided into three regions:

- *Reactive near-field region.* This region is the area of space immediately surrounding the antenna in which the reactive components predominate. The size of the region varies, depending on the antenna design. For most antennas, the reactive near-field region extends two wavelengths or less from the radiating elements.
- *Radiating near-field region.* This region is characterized by the predictable distribution of the radiating field. In the near-field region, the relative angular distribution of the field is dependent on the distance from the antenna.
- *Far-field region.* This region is characterized by the independence of the relative angular distribution of the field with varying distance. The pattern is essentially independent of distance.

12.1.2   Impedance Matching

Most practical antennas require some form of impedance matching between the transmission line and the radiating elements. The implementation of a matching network can take on many forms, depending upon the operating frequency and output power.

The *negative sign* convention is generally used in impedance matching analysis. That is, if a network delays or retards a signal by θ degrees, the phase shift across the network is said to be minus θ degrees. For example, a 1/4-wavelength of transmission line, if properly terminated, has a phase shift of -90°. Thus, a *lagging* or low-pass network has a negative phase shift, and a *leading* or high-pass network has a positive phase shift. There are three basic network types that can be used for impedance matching: L, pi, and tee.

L Network. The L network is shown in Figure 12-2. The loaded Q of the network is determined from equation 1. Equation 2 defines the shunt leg reactance, which θ is negative (capacitive) when θ is negative, and positive (inductive) when θ is positive. The series leg reactance is found using equation 3, the phase shift via equation 4, and the currents and voltages via Ohm's Law. Note that R_2 (the resistance on the shunt leg side of the L network) must always be greater than R_1. An L network cannot be used to match equal resistances or adjust phase independently of resistance.

Tee Network. The tee network is shown in Figure 12-3. This configuration can be used to match unequal resistances. The tee network has the added feature that phase shift is independent of the resistance transformation ratio. A tee network can be considered simply as two L networks back-to-back. Note that there are two loaded Qs associated with a tee network — an input Q and output Q. In order to gauge the bandwidth of the tee network, the lower value of Q must be ignored. Note that the Q of a tee network increases with increasing phase shift.

Equations 5 through 14 describe the tee network. It is a simple matter to find the input and output currents via Ohm's Law, and the shunt leg current can be found via the Cosine Law (equation 12). Note that this current increases with increasing phase shift. Equation 13 describes the mid-point resistance of a tee network, which is always higher than R_1 or R_2. Equation 14 is useful when designing a *phantom tee network*; that is, where X_2 is made up only of the antenna reactance, and there is no physical component in place of X_2. Keep in mind that a tee network is considered as having a lagging or negative phase shift when the shunt leg is capacitive (X_3 negative),

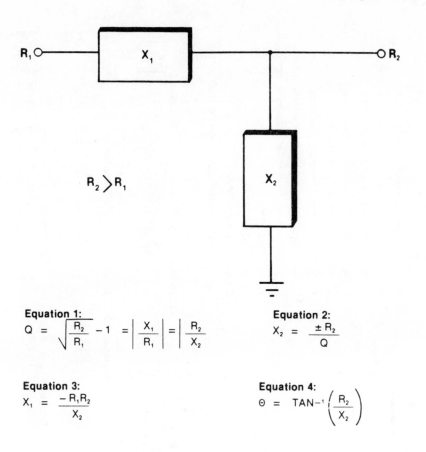

Equation 1:
$$Q = \sqrt{\frac{R_2}{R_1} - 1} = \left| \frac{X_1}{R_1} \right| = \left| \frac{R_2}{X_2} \right|$$

Equation 2:
$$X_2 = \frac{\pm R_2}{Q}$$

Equation 3:
$$X_1 = \frac{-R_1 R_2}{X_2}$$

Equation 4:
$$\Theta = TAN^{-1}\left(\frac{R_2}{X_2} \right)$$

Where: R_1 = L network input resistance (ohms)

R_2 = L network output resistance (ohms)

X_1 = Series leg reactance (ohms)

X_2 = Shunt leg reactance (ohms)

Q = Loaded Q of the L network

Figure 12-2 L network parameters. (*Source: Broadcast Engineering Magazine.*)

and vice versa. The input and output arms can be either negative or positive, depending on the resistance transformation ratio and desired phase shift.

Pi Network. The pi network is shown in Figure 12-4. It can also be considered as two L networks back-to-back and, therefore, the

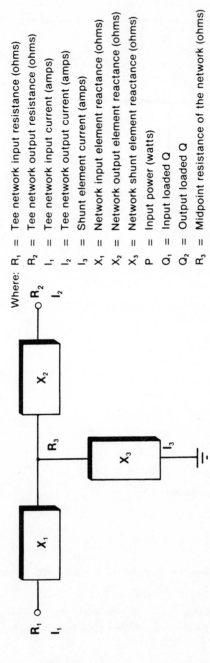

Where: R_1 = Tee network input resistance (ohms)
R_2 = Tee network output resistance (ohms)
I_1 = Tee network input current (amps)
I_2 = Tee network output current (amps)
I_3 = Shunt element current (amps)
X_1 = Network input element reactance (ohms)
X_2 = Network output element reactance (ohms)
X_3 = Network shunt element reactance (ohms)
P = Input power (watts)
Q_1 = Input loaded Q
Q_2 = Output loaded Q
R_3 = Midpoint resistance of the network (ohms)

Equation 5:
$$X_3 = \frac{\sqrt{R_1 R_2}}{\text{SIN}(\Theta)}$$

Equation 6:
$$X_1 = \frac{R_1}{\text{TAN}(\Theta)} - X_3$$

Equation 7:
$$X_2 = \frac{R_2}{\text{TAN}(\Theta)} - X_3$$

Equation 8:
$$Q_1 = \left| \frac{X_1}{R_1} \right|$$

Equation 9:
$$Q_2 = \left| \frac{X_2}{R_2} \right|$$

Equation 10:
$$I_1 = \sqrt{\frac{P}{R_1}}$$

Equation 11:
$$I_2 = \sqrt{\frac{P}{R_2}}$$

Equation 12:
$$I_3 = \sqrt{I_1^2 + I_2^2 - 2(I_1)(I_2)\text{COS}(\Theta)}$$

Equation 13:
$$R_3 = (Q_2^2 + 1)R_2$$

Equation 14:
$$\Theta = \text{TAN}^{-1}\left(\frac{X_1}{R_1}\right) \pm \text{TAN}^{-1}\left(\frac{X_2}{R_2}\right)$$

Figure 12-3 Tee network parameters. (Source: Broadcast Engineering Magazine.)

same comments about overall loaded Q apply. Note that suscep-
tances have been used in equations 15 through 19 instead of reactances in order to simplify calculations. The same conventions re-
garding tee network currents apply to pi network voltages (equations
20, 21, and 22). The mid-point resistance of a pi network is always
less than R_1 or R_2. A pi network is considered as having a negative
or lagging phase shift when Y_3 is positive, and vice versa.

Line Stretcher. A *line stretcher* makes a transmission line look
longer or shorter in order to produce sideband impedance symmetry
at the transmitter PA (see Figure 12-5). This is done to reduce audio
distortion in an envelope detector — the kind of detector that most
AM receivers employ. Symmetry is defined as equal sideband resistances, and equal — but opposite sign — sideband reactances. There
are two possible points of symmetry, each 90° from the other. One
produces sideband resistances greater than the carrier resistance,
and the other produces the opposite effect. One side will create a
pre-emphasis effect, and the other a de-emphasis effect. Depending
on the Q of the transmitter output network, one point of symmetry
may yield lower sideband VSWR at the PA than the other. This results from the Q of the output network opposing the Q of the antenna in one direction, but aiding the antenna Q in the other direction.

12.1.3 Antenna Types

The *dipole antenna* is simplest of all antennas, and the building
block of most other designs. The dipole consists of two in-line rods or
wires with a total length equal to 1/2-wave at the operating frequency. Figure 12-6 shows the typical configuration, with two 1/4-
wave elements connected to a transmission line. The radiation resistance of a dipole is on the order of 73Ω. The bandwidth of the antenna may be increased by increasing the diameter of the elements,
or by using cones or cylinders rather than wires or rods, as shown in
the figure. Such modifications also increase the impedance of the antenna.

The dipole can be straight (in-line) or bent into a V-shape. The
impedance of the V-dipole is a function of the V angle. Changing the
angle effectively tunes the antenna. The vertical radiation pattern of
the V-dipole antenna is similar to the straight dipole for angles of
120° or less.

A *folded dipole* may be fashioned as shown in Figure 12-7. Such a
configuration results in increased bandwidth and impedance. Imped-

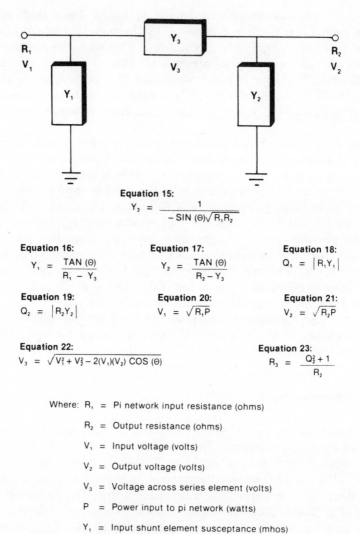

Equation 15:
$$Y_3 = \frac{1}{-\,SIN\,(\Theta)\sqrt{R_1R_2}}$$

Equation 16:
$$Y_1 = \frac{TAN\,(\Theta)}{R_1 - Y_3}$$

Equation 17:
$$Y_2 = \frac{TAN\,(\Theta)}{R_2 - Y_3}$$

Equation 18:
$$Q_1 = |R_1Y_1|$$

Equation 19:
$$Q_2 = |R_2Y_2|$$

Equation 20:
$$V_1 = \sqrt{R_1P}$$

Equation 21:
$$V_2 = \sqrt{R_2P}$$

Equation 22:
$$V_3 = \sqrt{V_1^2 + V_2^2 - 2(V_1)(V_2)\,COS\,(\Theta)}$$

Equation 23:
$$R_3 = \frac{Q_2^2 + 1}{R_2}$$

Where: R_1 = Pi network input resistance (ohms)

R_2 = Output resistance (ohms)

V_1 = Input voltage (volts)

V_2 = Output voltage (volts)

V_3 = Voltage across series element (volts)

P = Power input to pi network (watts)

Y_1 = Input shunt element susceptance (mhos)

Y_2 = Output shunt element susceptance (mhos)

Y_3 = Series element susceptance (mhos)

Q_1 = Input loaded Q

Q_2 = Output loaded Q

Figure 12-4 Pi network parameters. (*Source: Broadcast Engineering Magazine.*)

LINE STRETCHER CONFIGURATION

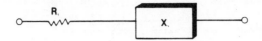

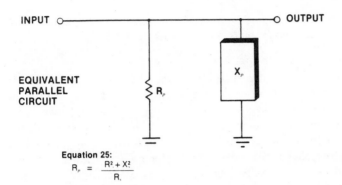

INPUT

OUTPUT

EQUIVALENT
PARALLEL
CIRCUIT

X_p

R_p

Equation 25:
$$R_p = \frac{R_s^2 + X_s^2}{R_s}$$

Equation 26:
$$X_p = \frac{R_s^2 + X_s^2}{X_s}$$

Where: R_s = Series configuration resistance (ohms)

R_p = Parallel configuration resistance (ohms)

X_s = Series reactance (ohms)

X_p = Parallel reactance (ohms)

Figure 12-5 Line stretcher configuration. (*Source: Broadcast Engineering Magazine.*)

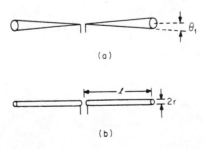

(a)

(b)

Figure 12-6 Half-wave dipole antenna: (a) conical dipole; (b) conventional dipole. (*Source:* K. Blair Benson, *Television Engineering Handbook,* McGraw-Hill, New York, 1986.)

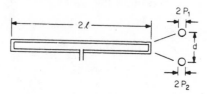

Figure 12-7 Folded dipole antenna. (*Source:* K. Blair Benson, *Television Engineering Handbook,* McGraw-Hill, New York, 1986.)

ance can be further increased by using rods of different diameter and by varying the spacing of the elements. The 1/4-wave dipole elements connected to the closely-coupled 1/2-wave element act as a matching stub between the transmission line and the single-piece 1/2-wave element. This broadbands the folded dipole antenna by a factor of two.

A *corner-reflector* antenna may be formed as shown in Figure 12-8. A ground plane or flat reflecting sheet is placed at a distance of 1/16- to 1/4-wavelengths behind the dipole. Gain in the forward direction may be increased by a factor of two with this type of design.

Quarter-Wave Monopole. A conductor placed above a ground plane forms an image in the ground plane such that the resulting pattern is a composite of the *real antenna* and the *image antenna* (see Figure 12-9). The operating impedance is 1/2 of the impedance of the antenna and its image when fed as a physical antenna in free space. An example will help illustrate this concept. A 1/2-wave monopole mounted on an infinite ground plane has an impedance equal to 1/2 the free-space impedance of a 1/4-wave dipole. It follows, then, that the theoretical characteristic resistance of a 1/4-wave monopole with an infinite ground plane is 37Ω.

For a real-world antenna, an infinite ground plane is neither possible nor required. An antenna mounted on a ground plane that is two

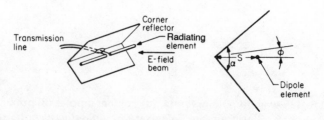

Figure 12-8 Corner-reflector antenna. (*Source:* K. Blair Benson, *Television Engineering Handbook,* McGraw-Hill, New York, 1986.)

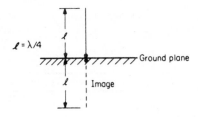

Figure 12-9 Vertical monopole mounted above a ground plane. (*Source:* K. Blair Benson, *Television Engineering Handbook,* McGraw-Hill, New York, 1986.)

to three times the operating wavelength has about the same impedance as a similar antenna mounted on an infinite ground plane.

Log-Periodic Antenna. The log-periodic antenna can take on a number of forms. Typical designs include the *conical log spiral, log-periodic V,* and *log-periodic dipole.* The most common of these antennas is the log-periodic dipole. The antenna can be fed either by using alternating connections to a balanced line, or by a coaxial line running through one of the feeders from front to back. In theory, the log-periodic antenna may be designed to operate over many octaves. In practice, however, the upper frequency is limited by the precision required in constructing the small elements, feed lines, and support structure of the antenna.

Yagi-Uda Antenna. The Yagi-Uda is an *end-fire array* consisting typically of a single driven dipole with a *reflector dipole* behind the driven element, and one or more *parasitic director elements* in front (see Figure 12-10). Common designs use from one to seven director elements. As the number of elements increases, so does directivity.

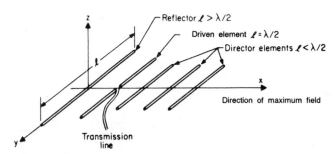

Figure 12-10 The Yagi-Uda array. (*Source:* K. Blair Benson, *Television Engineering Handbook,* McGraw-Hill, New York, 1986.)

Table 12-1 Typical characteristics of single-channel Yagi-Uda arrays.
(*Source:* K. Blair Benson, *Television Engineering Handbook,* McGraw-Hill, New York, 1986.)

No. of elements	Gain, dB	Beam width, deg
2	3–4	65
3	6–8	55
4	7–10	50
5	9–11	45
9	12–14	37
15	14–16	30

Bandwidth, however, decreases as the number of elements increase. Arrays of more than four director elements are typically narrowband.

The radiating element is 1/2-wavelength at the center of the band covered. The single reflector element is slightly longer, and the director elements are slightly shorter, all spaced approximately 1/4-wavelength from each other.

Table 12-1 demonstrates how the number of elements determines the gain and beamwidth of a Yagi-Uda antenna.

Waveguide Antenna. The waveguide antenna consists of a dominant-mode-fed waveguide opening onto a conducting ground plane. Designs may be based on rectangular, circular or coaxial waveguide (also called an *annular slot*). The slot antenna is simplicity itself. A number of holes of a given dimension are placed at intervals along a section of waveguide. The radiation characteristics of the antenna are determined by the size, location, and orientation of the slots. The antenna offers optimum reliability because there are no discrete elements, except for the waveguide section itself.

Horn Antenna. The horn antenna may be considered a natural extension of the dominant-mode waveguide feeding the horn in a manner similar to the wire antenna, which is a natural extension of the two-wire transmission line. The most common types of horns are the *E-plane sectoral, H-plane sectoral, pyramidal horn* (formed by expanding the walls of the $TE_{0,1}$-mode-fed rectangular waveguide) and the *conical horn* (formed by expanding the walls of the $TE_{1,1}$-mode-fed circular waveguide). The H-plane sectoral horn is shown in Fig-

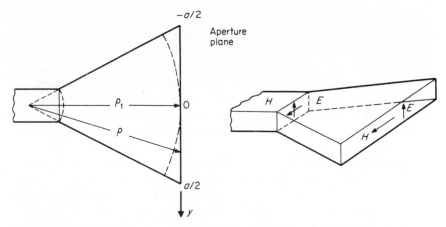

Figure 12-11 Geometry of an H-plane sectoral horn. (From Compton and Collin in Fink and Christiansen, *Electronics Engineers' Handbook*, McGraw-Hill, New York, 1989.)

ure 12-11. Dielectric-loaded waveguides and horns offer improved pattern performance over unloaded horns. Ridged and tapered horn designs improve the bandwidth characteristics. Horn antennas are available in single and dual polarized configurations.

Reflector Antenna. The reflector antenna is formed by mounting a radiating feed antenna above a reflecting ground plane. The most basic form of reflector is the loop or dipole spaced over a finite ground plane. This concept is the basis for the parabolic or spherical reflector antenna. The *parabolic reflector antenna* may be fed directly, or through the use of a subreflector in the focal region of the parabola. In this approach, the subreflector is illuminated from the parabolic surface. The chief disadvantage of this design is the aperture blockage of the subreflector, which restricts its use to large aperture antennas. The operation of a parabolic or spherical reflector antenna is typically described using physical optics.

Parabolic-reflector antennas are usually illuminated by a flared-horn antenna with a flare angle of less than 18°. A rectangular horn with a flare angle of less than 18° has approximately the same aperture field as the dominant-mode rectangular waveguide feeding the horn.

Spiral Antenna. The bandwidth limitations of an antenna are based on the natural change in the critical dimensions of the radiating elements caused by differing wavelengths. The spiral antenna

overcomes this limitation because the radiating elements are specified only in angles. A two-arm equiangular spiral is shown in Figure
12-12. This common design gives wideband performance. Circular polarization is inherent in the antenna. Rotation of the pattern corresponds to the direction of the spiral arms. Typically, the gain of a
spiral antenna is slightly higher than a dipole. The basic spiral antenna radiates on both sides of the arms. Unidirectional radiation is
achieved through the addition of a reflector or cavity.

Array Antenna. The term "array antenna" covers a wide variety
of physical structures. The most common configuration is the *planar
array antenna,* which consists of a number of radiating elements reg-
ularly spaced on a rectangular or triangular lattice. The *linear array
antenna,* where the radiating elements are placed in a single line, is
also common. The pattern of the array is the product of the element
pattern and the array configuration. Large array antennas may consist of 20 or more radiating elements.

Correct phasing of radiating elements is the key to the operation of
the system. The radiating pattern of the structure, including direction, may be controlled through proper adjustment of the relative
phase of the elements.

Figure 12-12 Two-arm equiangular spiral antenna.

12.1.4   Antenna Applications

An analysis of the applications of antennas for commercial and industrial use is beyond the scope of this chapter. It is instructive, however, to examine three of the most obvious antenna applications: AM and FM radio, and television. These applications illustrate antenna technology as it applies to frequencies ranging from the lower end of the MF band to the upper reaches of UHF.

12.2 AM Broadcast Antenna Systems

Vertical polarization of the transmitted signal is used for AM broadcast stations because of its superior groundwave propagation, and because of the simple antenna designs that it affords. The FCC and licensing authorities in other countries have established classifications of AM stations specifying power levels and hours of operation. Protection requirements set forth by the FCC specify that some AM stations (in the U.S.) reduce their transmitter power at sunset and return to full power at sunrise. This method of operation is based on the propagation characteristics of AM band frequencies. AM signals propagate further at nighttime than during the day.

The different day/night operating powers are designed to provide each AM station with a specified coverage area that is free from interference. Theory rarely translates into practice insofar as coverage is concerned, however, because of the increased interference that all AM stations suffer at nighttime.

The tower you see at any AM radio station transmitter site is only half of the antenna system. The second element is a buried ground system. Current on a tower does not simply disappear, rather it returns to earth through the capacitance between the earth and the tower. Ground losses are greatly reduced if the tower has a radial copper ground system. A typical single tower ground system is made up of 120 radial ground wires, each 140 electrical degrees long (at the operating frequency) equally spaced out from the tower base. This is often augmented with an additional 120 interspersed radials 50 ft. long.

12.2.1   Directional AM Antenna Design

When a non-directional antenna with a given power does not radiate enough energy to serve the station's primary service area, or radiates too much energy toward other radio stations on the same or adjacent frequencies, it is necessary to employ a directional antenna system. Rules set out by the FCC and regulatory agencies in other countries specify the protection requirements to be provided by various classes of stations, for both daytime and nighttime hours. These limits tend to define the shape and size of the most desirable antenna pattern.

A directional antenna functions by carefully controlling the amplitude and phase of the RF currents fed to each tower in the system. The directional pattern is a function of the number and spacing of the towers (vertical radiators), and the relative phase and magnitude of their currents. The number of towers in a directional AM array can range from two to six or even more in a complex system. One tower is defined as the *reference tower*. The amplitude and phase of the other towers are measured relative to this reference.

A complex network of power splitting, phasing, and antenna coupling elements is required to make a directional system work. Figure 12-13 shows a block diagram of a basic two-tower array. A power divider network controls the relative current amplitude in each tower. A phasing network provides control of the phase of each tower current, relative to the reference tower. Matching networks at the base of each tower couple the transmission line impedance to the base operating impedance of the radiating towers.

In practice, the system shown in the figure would not consist of individual elements. Instead, the matching network, power dividing network, and phasing network would all usually be combined into a single unit, referred to as the *phasor*.

12.2.2   Antenna Pattern Design

The pattern of any AM directional antenna system (array) is determined by a number of factors, including:

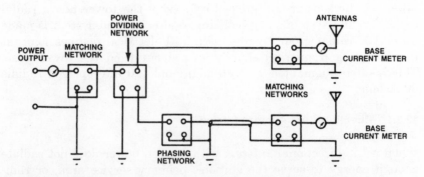

Figure 12-13 Block diagram of an AM directional antenna feeder system for a two-tower array. (*Source:* Benson and Whitaker, *Television and Audio Handbook,* McGraw-Hill, New York, 1990.)

- Electrical parameters (phase relationship and current ratio for each tower).
- Height of each tower.
- Position of each tower with respect to the other towers (particularly with respect to the reference tower).

A *directional array* consists of two or more towers arranged in a specific manner on a plot of land. Figure 12-14 shows a typical 3-tower array, and the pattern such an array would produce. This is an *in-line array,* meaning that all the elements (towers) are in line with one another. Notice that the *major lobe* is centered on the same line as the line of towers, and that the *pattern nulls (minima)* are

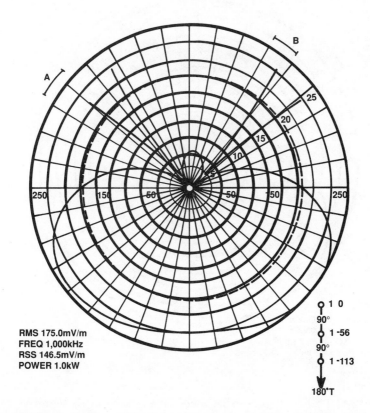

Figure 12-14 Radiation pattern generated with a 3-tower in-line directional array using the electrical parameters and orientation shown. (*Source: Broadcast Engineering Magazine.*)

positioned symmetrically about the line of towers, protecting co-channel stations A and B at true bearings of 315° and 45° respectively.

Figure 12-15 shows the same array, except that it has been rotated by 10°. Notice that the pattern shape is not changed, but the position of the major lobe and the nulls follow the line of towers. Also notice that the nulls are no longer pointed at the stations to be protected. Figure 12-16 and Figure 12-17 show that the position of the nulls can be changed by varying the electrical parameters so that one or the other can be pointed in the required direction, but not both. Also, when this is attempted, the size and shape of the small *back lobe* is changed, as is the shape of the major lobe, especially the radiation on the line of towers.

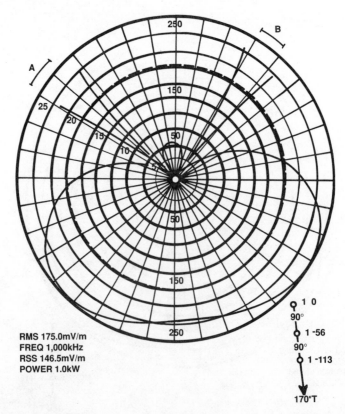

Figure 12-15 Radiation pattern produced when the array of Figure 12-14 is rotated to a new orientation. (*Source: Broadcast Engineering Magazine.*)

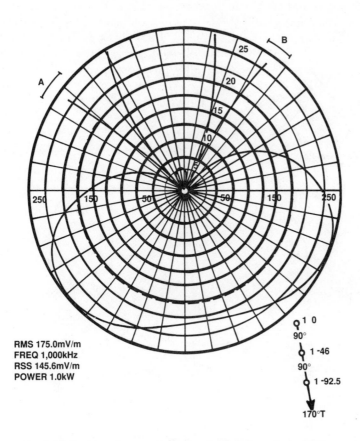

RMS 175.0mV/m
FREQ 1,000kHz
RSS 145.6mV/m
POWER 1.0kW

1 0
90°

1 -46
90°

1 -92.5

170°T

Figure 12-16 Directional pattern generated with the orientation shown in Figure 12-15, but with different electrical parameters. (*Source: Broadcast Engineering Magazine.*)

If this directional antenna system were constructed on a gigantic turntable, the pattern could be rotated without affecting the shape. But, to accomplish the required protections and to have the major lobe(s) oriented in the right direction, there is only one correct position. In most cases, the position of the towers will be specified with respect to a single reference tower. The location of the other towers will be given in the form of distance and bearing from that reference. Occasionally, a reference point, usually the center of the array, will be used as a geographic coordinate point.

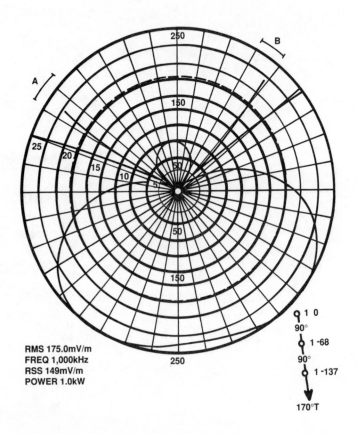

RMS 175.0mV/m
FREQ 1,000kHz
RSS 149mV/m
POWER 1.0kW

Figure 12-17 Directional pattern generated by the tower orientation shown in Figure 12-15, but with a new set of electrical parameters. (*Source: Broadcast Engineering Magazine.*)

Bearing. The bearing or azimuth of the towers from the reference tower or point is specified clockwise in degrees from *true north*. The distinction between true and *magnetic north* is vital. The magnetic North Pole is not the true or geographic North Pole. (In fact, it is in the vicinity of 74° north, 101° west, in the islands of northern Canada.) The difference between magnetic and true bearings is called variation or *magnetic declination*. Declination, a term generally used by surveyors, varies for different locations. It is not a constant. The earth's magnetic field is subject to a number of changes in intensity and direction. These changes take place over daily, yearly, and long-term (or *secular*) periods. The secular changes result in a relatively constant increase or decrease in declination over a period of many years.

12.2.3 Antenna Monitoring System

Monitoring the operation of an AM directional antenna basically involves measuring the power into the system, the relative value of currents into the towers, their phase relationships, and the levels of radiated signal at certain monitoring points some distance from the antenna. Figure 12-18 shows a block diagram of a typical monitoring system for a three-tower array. For systems with additional towers, the basic layout is extended by adding more pickup elements, sample lines, and a monitor with additional inputs.

Phase/Current Sample. Two types of phase/current sample pickup elements are commonly used — the *sample loop* and *torroidal current transformer* (TCT). The sample loop consists of a single turn unshielded loop of rigid construction, with a fixed gap at the open end for connection of the sample line. The device must be mounted on the tower near the point of maximum current. The loop may be used on towers of both uniform and non-uniform cross section. It must operate at tower potential, except for towers of less than 130° electrical height, where the loop may be operated at ground potential. Figure 12-19 shows a typical sample loop, with insulators for mounting at ground potential, where permitted.

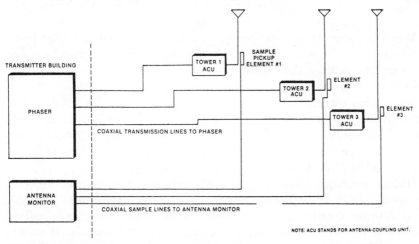

Figure 12-18 A typical 3-tower directional antenna monitoring system. (*Source: Broadcast Engineering Magazine.*)

Figure 12-19 A fixed, non-shielded RF sampling loop. (Courtesy of Harris Corp.)

When the sample loop is operated at tower potential, the coax from the loop to the base of the tower is also at tower potential. In order to bring the sample line across the base of the tower, a sample line isolation coil is used.

A shielded torroidal current transformer may also be used as the phase/current pickup element. Such devices offer several advantages over the sample loop including greater stability and reliability. Because they are located inside the tuning unit cabinet or house, TCTs are protected from wind, rain, ice, and vandalism.

Unlike the rigid, fixed sample loop, torroidal current transformers are available in several sensitivities, ranging from 0.25-1.0 V per ampere of tower current. Tower currents of up to 40 A may be handled, providing a more usable range of voltages for the antenna monitor. Figure 12-20 shows a typical TCT. Figure 12-21 shows the various

Figure 12-20 A torroidal current transformer for AM applications. (Courtesy of Delta Electronics.)

arrangements that may be used for phase/current sample pickup elements.

Sample lines. The selection and installation of the sampling lines for a directional monitoring system are important factors in the ultimate accuracy of the overall array.

With *critical arrays* (antennas requiring operation within tight limits specified in the station license), all sample lines must be of equal electrical length and installed in such a manner that the corresponding lengths of all lines are exposed to equal environmental conditions.

While sample lines may be run above ground on supports (if protected and properly grounded) the most desirable arrangement is direct burial using jacketed cable. Burial of sample line cable is almost a standard practice because proper burial offers good protection against physical damage and a more stable temperature environment.

The Common Point. The power input to a directional antenna is measured at the phasor *common point*. Power is determined by the *direct method*:

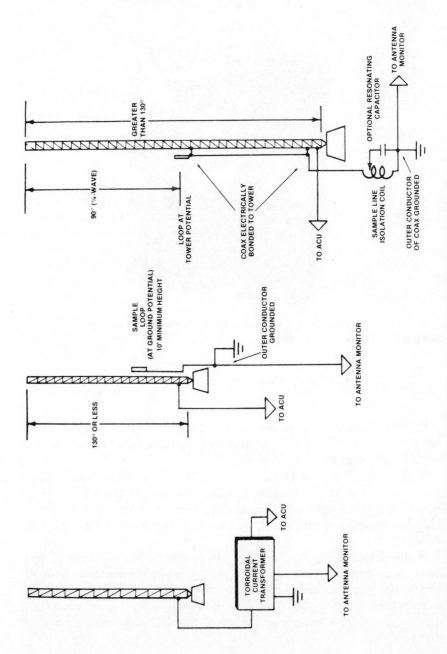

Figure 12-21 Three possible circuit configurations for phase sample pickup. (Source: Broadcast Engineering Magazine.)

$$P = I^2 R$$

Where:
P = power in W
I = the common point current in Amps
R = the common point resistance in Ω

Monitor Points. Routine monitoring of a directional antenna involves measuring field intensity at certain locations away from the
antenna called *monitor points*. These points are selected and established during the initial tune-up of the antenna system. Measurements at monitor points should confirm that radiation in prescribed
directions does not exceed a value that would cause interference to
other stations operating on the same or adjacent frequencies. The
field intensity limits at these points are normally specified in the
station license. Measurements at the monitor points may be required
on a weekly or a monthly basis, depending on several factors and
conditions relating to the particular station. If the system is not a
critical array, quarterly measurements may be sufficient.

12.2.4 Folded Unipole Antenna

The *folded unipole* antenna consists of a grounded vertical structure
with one or more conductors folded back parallel to the side of the
structure. It can be visualized as a half-wave folded dipole perpendicular to the ground and cut in half (see Figure 12-22). This design
makes it possible to provide a wide range of resonant radiation resistances by varying the ratio of the diameter of the folded-back conductors in relation to the tower. Top loading can also be used to

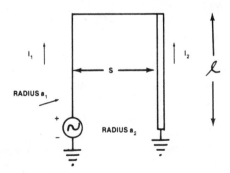

**Figure 12-22 The folded unipole antenna can be thought of as a 1/4-wave
folded dipole antenna perpendicular to the ground and cut in half.
(*Source: Broadcast Engineering Magazine.*)**

broaden the antenna bandwidth. A side view of the folded unipole is shown in Figure 12-23.

The folded unipole antenna could be considered a modification of the standard shunt-fed system. Instead of a slant wire that leaves the tower at an approximate 45° angle (as used for shunt-fed systems), the folded unipole antenna has one or more wires attached to the tower at a predetermined height. The wires are supported by standoff insulators and run parallel to the sides of the tower down to the base.

The tower is grounded at the base. The folds, or wires, are joined together at the base and driven through an impedance matching network. Depending upon the tuning requirements of the folded unipole, the wires may be connected to the tower at the top and/or at predetermined levels along the tower with shorting stubs.

The folded unipole can be used on tall (130° or greater) towers. However, if the unipole is not divided into two parts, the overall efficiency (unattenuated field intensity) will be considerably lower than the normally expected field for the electrical height of the tower.

12.3 FM Broadcast Antenna Systems

The propagation characteristics of VHF FM radio are much different than for MF AM. There is essentially no difference between day and night FM propagation. FM stations have relatively uniform day and night service areas with the same operating power.

A wide variety of antennas is available for use in the FM broadcast band. Nearly all employ circular polarization. Although antenna designs differ from one manufacturer to another, generalizations can be made that apply to most units.

12.3.1   Antenna Types

There are three basic classes of FM broadcast transmitting antennas in use today: *ring stub* and *twisted ring*, *shunt-* and *series-fed slanted dipole*, and *multi-arm short helix*. While each design is unique, all have the following items in common:

- The antennas are designed for side mounting to a steel tower or pole.
- Radiating elements are shunted across a common rigid coaxial transmission line.
- Elements are placed along the rigid line every one wavelength.

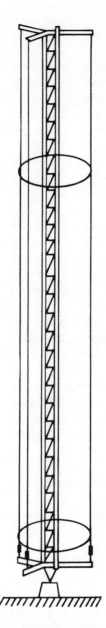

Figure 12-23 The folds of the unipole antenna are arranged either near the legs of the tower or near the faces of the tower. (*Source: Broadcast Engineering Magazine.*)

- Antennas with one to seven bays are fed from the bottom of the coaxial transmission line.
- Antennas with more than seven bays are fed from the center of the array to provide more predictable performance in the field.
- Antennas generally include a means of tuning out reactances after the antenna has been installed through the adjustment of variable capacitive or inductive elements at the feed point.

Figure 12-24 shows a shunt-fed slanted dipole antenna that consists of two half-wave dipoles offset 90°. The two sets of dipoles are rotated 22.5° (from their normal plane) and are *delta-matched* to provide a 50 Ω impedance at the radiator input flange. The lengths of all four dipole arms may be matched to resonance by mechanical adjustment of the end fittings. Shunt-feeding (when properly adjusted) provides equal currents in all four arms.

Wideband *panel antennas* are a fourth common type of antenna used for FM broadcasting. Panel designs share some of the characteristics listed previously, but are intended primarily for specialized installations in which two or more stations will use the antenna simultaneously. Panel antennas are larger and more complex than other FM antennas, but offer the possibility for shared tower space among several stations, and custom coverage patterns that would be difficult or even impossible with the more common designs. The ideal combination of antenna gain and transmitter power for a particular installation involves the analysis of a number of parameters. As shown in Table 12-2, a variety of pairings can be made to achieve the same ERP.

12.4 Television Antenna Systems

Television broadcasting uses horizontal polarization for the majority of installations worldwide. More recently, interest in the advantages of circular polarization has resulted in an increase in this form of transmission, particularly for VHF channels.

Both horizontal and circular polarization designs are suitable for tower-top or side-mounted installations. The latter option is dictated primarily by the existence of a previously-installed tower-top antenna. On the other hand, in metropolitan areas where several antennas must be located on the same structure, either a stacking or candelabra-type arrangement is feasible. Figure 12-25 shows an example of antenna stacking on the top of the John Hancock Center in Chicago, where six TV antennas, four standby antennas, and numerous FM transmitting antennas are located. Figure 12-26 shows a

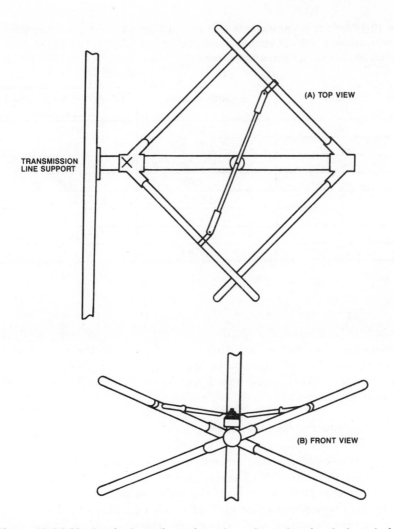

TRANSMISSION LINE SUPPORT

(A) TOP VIEW

(B) FRONT VIEW

Figure 12-24 Mechanical configuration of one bay of a circularly polarized FM transmitting antenna (Jampro JSCP series): (a) top view of the antenna; (b) front view. (*Source:* Benson and Whitaker, *Television and Audio Handbook,* McGraw-Hill, New York, 1990.)

candelabra installation atop the Mt. Sutro tower in San Francisco. The Sutro tower supports on its uppermost level eight TV antennas. A number of FM transmitting antennas and two-way radio antennas are located on lower levels of the structure.

Another approach to TV transmission involves combining the RF outputs of two stations and feeding a single wideband antenna. This

Table 12-2 Various combinations of transmitter power and antenna gain that will produce 100 kW effective radiated power (ERP) for an FM station. (*Source: Broadcast Engineering Magazine.*)

No. bays	Antenna gain	Transmitter power* (Kilowatts)
3	1.5888	66.3
4	2.1332	49.3
5	2.7154	38.8
6	3.3028	31.8
7	3.8935	27.0
8	4.4872	23.5
10	5.6800	18.5
12	6.8781	15.3

approach is expensive and requires considerable engineering analysis to produce a combiner system that will not degrade the performance of either transmission system. In the Mt. Sutro example, it can be seen that two stations (channels 4 and 5) are combined into a single transmitting antenna.

12.4.1 Top-Mounted Antenna Types

The typical television broadcast antenna is a broadband radiator operating over a bandwidth of several megahertz with an efficiency of over 95 percent. Reflections from the antenna and transmission line back to the transmitter must be kept small enough to introduce negligible picture degradation. Furthermore, the gain and pattern characteristics of the antenna must be designed to achieve the desired coverage within acceptable tolerances. Tower-top, pole-type antennas designed to meet these parameters can be classified into two categories — *resonant dipoles* and *multi-wavelength traveling-wave elements*.

The primary considerations in the design of a top-mounted antenna are the achievement of uniform omnidirectional azimuth fields and minimum windloading. A number of different approaches have been tried successfully. Figure 12-27 illustrates the basic mechanical design of the most common antennas.

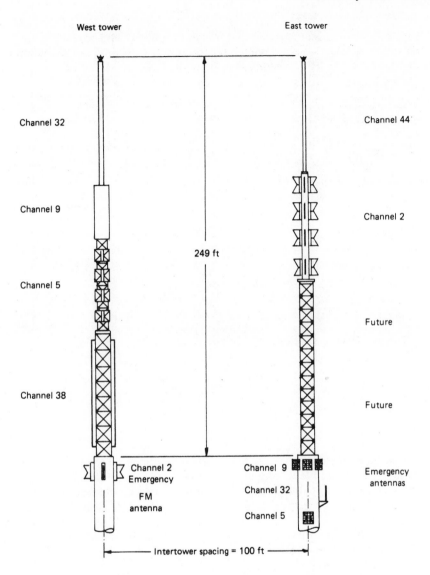

West tower

East tower

Channel 32

Channel 44

Channel 9

Channel 2

249 ft

Channel 5

Channel 38

Future

Future

Channel 2
Emergency

Channel 9

Emergency
antennas

FM
antenna

Channel 32

Channel 5

Intertower spacing = 100 ft

Figure 12-25 Twin tower antenna array atop the John Hancock Center in Chicago. Note how antennas have been stacked to overcome space restrictions. (*Source:* Benson and Whitaker, *Television and Audio Handbook,* McGraw-Hill, New York, 1990.)

Turnstile Antenna. The turnstile is the earliest and most popular resonant antenna for VHF broadcasting. This antenna is made up of four *batwing*-shaped elements mounted on a vertical pole in a manner resembling a turnstile. The four batwings are, in effect, two

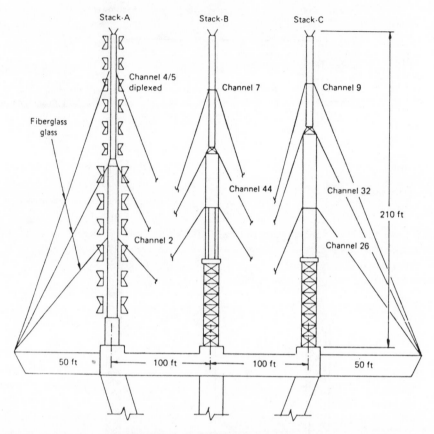

Figure 12-26 Installation of TV transmitting antennas on the candelabra structure at the top level of the Mt. Sutro tower in San Francisco. This installation makes extensive use of antenna stacking. (*Source:* K. Blair Benson, *Television Engineering Handbook,* McGraw-Hill, New York, 1986.)

dipoles fed in quadrature phase. The azimuth-field pattern is a function of the diameter of the support mast. The pattern is usually within 10-15 percent of a true circle.

The turnstile antenna is made up of several layers, usually six for channels 2 through 6 and twelve for channels 7 through 13. The turnstile is not suitable for side-mounting, except for standby applications in which coverage degradation can be tolerated.

Coax Slot Antenna. Commonly referred as the *pylon antenna,* the coax slot is the most popular top-mounted unit for UHF applications. Horizontally polarized radiation is achieved by using axial resonant slots on a cylinder to generate RF current around the outer

Turnstile	Coax slot	Waveguide slot	Zig zag	Helix	Multislot traveling wave

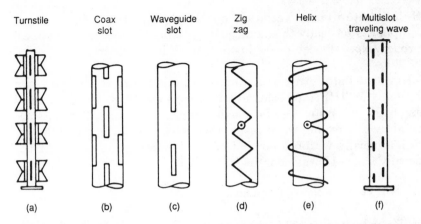

(a)	(b)	(c)	(d)	(e)	(f)

Figure 12-27 Various antennas used for VHF and UHF broadcasting. All designs provide linear (horizontal) polarization. Illustrated are: (a) turnstile antenna; (b) coax slot antenna; (c) waveguide slot antenna; (d) zigzag antenna; (e) helix antenna; (f) multi-slot traveling-wave antenna. (*Source: Benson and Whitaker, Television and Audio Handbook, McGraw-Hill, New York, 1990.*)

surface of the cylinder. A good omnidirectional pattern is achieved by exciting four columns of slots around the circumference, which is basically just a section of rigid coaxial transmission line. The slots along the pole are spaced approximately one wavelength per layer and a suitable number of layers are used to achieve the desired gain. Typical gains range from 20-40. By varying the number of slots around the periphery of the cylinder, directional azimuth patterns can be achieved.

Waveguide Slot Antenna. The UHF waveguide slot is a variation on the coax slot antenna. The antenna is simply a section of waveguide with slots cut into the sides. The physics behind the design is long and complicated. However, the end result is the simplest of all antennas. This is a desirable feature in field applications because simple designs translate to long-term reliability.

Zigzag Antenna. The zigzag is a panel array design that utilizes a conductor routed up the sides of a 3- or 4-sided panel antenna in a zigzag manner. With this design, the vertical current component along the zigzag conductor is mostly canceled out, and the antenna can effectively be considered as an array of dipoles. With several such panels mounted around a polygonal periphery, the required azimuth pattern can be shaped by proper selection of feed currents to the various elements

Helix Antenna. A variation on the zigzag, the helix antenna accomplishes basically the same goal by using a different mechanical approach. Note the center feed point shown in the figure.

VHF Multi-slot Antenna. Similar mechanically to the coax slot antenna, the VHF multi-slot antenna consists of an array of axial slots on the outer surface of a coaxial transmission line. The slots are excited by a traveling wave inside the slotted line. The azimuth pattern is typically within 5 percent of omnidirectional. The antenna is generally about 15 wavelengths long.

12.4.2 Circularly Polarized Antennas

Circular polarization holds the promise of improved penetration into difficult coverage areas. There are a number of points to weigh in the decision of whether to use a circularly polarized (CP) antenna, however, not the least of which is that a station will have to double its transmitter power in order to maintain the same ERP. This assumes equal vertical and horizontal components.

It is possible, and sometimes desirable, to operate with *elliptical polarization*, in which the horizontal and vertical components are not equal. Further, the azimuth patterns for each polarization may be customized to provide the most efficient service area coverage.

Three major antenna types have been developed for CP television applications: the *normal mode helix*, various panel antenna designs, and the *interlaced traveling wave array*.

Normal Mode Helix. The normal mode helix consists of a supporting tube with helical radiators mounted around the tube on insulators. The antenna is broken into subarrays, each powered by a divider network. The antenna is called the "normal mode helix" because radiation occurs normal to the axis of the helix, or perpendicular to the support tube. The antenna provides an omnidirectional pattern.

Panel Antennas. The basic horizontally polarized panel antenna can be modified to produce circular polarization through the addition of vertically polarized radiators. Panel designs offer broad bandwidth and a wide choice of radiation patterns. By selecting the appropriate number of panels located around the tower, and the proper phase and amplitude distribution to the panels, a number of azimuth patterns may be realized. The primary drawback to the panel is the power distribution network required to feed the individual radiating elements.

Interlaced Traveling Wave Array. As the name implies, the radiating elements of this antenna are interlaced along an array, into which power is supplied. The energy input at the bottom of the antenna is extracted by the radiating elements as it moves toward the top. The antenna consists of a cylindrical tube that supports the radiating elements. The elements are slots for the horizontally polarized component, and dipoles for the vertically polarized component. The radiating elements couple RF directly off the main input line, so a power dividing network is not required.

12.4.3   Side-Mounted Antenna Types

Television antennas designed for mounting on the faces of a tower must meet the same basic requirements as a top-mounted antenna — wide bandwidth, high efficiency, predictable coverage pattern, high gain, and low wind loading — plus the additional challenge that the antenna must work in a less than ideal environment. Given the choice, no broadcaster would elect to place the transmitting antenna on the side of a tower instead of at the top. There are, however, a number of ways to solve the problems presented by side-mounting.

Butterfly Antenna. The butterfly is essentially a batwing panel developed from the turnstile radiator. The butterfly is one of the most popular panel antennas used for tower face applications. It is suitable for the entire range of VHF applications. A number of variations on the basic batwing theme have been produced, including modifying the shape of the turnstile-type wings to rhombus or diamond shapes. Another version utilizes multiple dipoles in front of a reflecting panel.

For CP applications, two crossed dipoles or a pair of horizontal and vertical dipoles are used. A variety of cavity-backed crossed-dipole radiators are also utilized for CP transmission.

The azimuth pattern of each panel antenna is unidirectional, and three or four such panels are mounted on the sides of a triangular or square tower to achieve an omnidirectional pattern. The panels can be fed in-phase, with each one centered on the face of the tower, or fed in rotating phase with the proper mechanical offset. In the latter case, the input impedance match is considerably better.

Directionalization of the azimuth pattern is realized by proper distribution of the feed currents to individual panels in the same layer. Stacking layers provides gains comparable with top-mounted antennas.

The main drawbacks of panel antennas are: (1) high wind-load, (2) complex feed system inside the antenna, and (3) restrictions on the size of the tower face, which determines to a large extent the omnidirectional pattern of the antenna.

UHF Side-Mounted Antennas. Utilization of panel antennas in a manner similar to those for VHF applications is not always possible at UHF installations. The high gains required for UHF broadcasting (in the range of 20-40, compared with gains of 6-12 for VHF) require far more panels with an associated complex feed system.

The zigzag panel antenna has been used for special omnidirectional and directional applications. For custom directional patterns, such as a cardioid shape, the pylon antenna can be side-mounted on one of the tower legs. Many stations, in fact, simply side-mount a pylon-type antenna on the leg of the tower that faces the greatest concentration of viewers. It is understood that viewers located behind the tower will receive a poorer signal, however, given the location of most TV transmitting towers — usually on the outskirts of their licensed city or on a mountain-top — this practice is often acceptable.

12.4.4   Broadband Antennas

Radiation of multiple channels from a single antenna requires the antenna to be broadband in both pattern and impedance (VSWR) characteristics. As a result, a broadband TV antenna represents a significant departure from the single channel pole antennas commonly used for VHF and UHF. The typical single channel UHF antenna uses a series feed to the individual radiating elements, while a broadband antenna has a branch feed arrangement. The two feed configurations are shown in Figure 12-28.

At the design frequency the series feed provides co-phased currents to its radiating elements. As the frequency varies, the electrical length of the series line feed changes so that the radiating elements are no longer in-phase outside of the designed channel. This electrical length change causes significant beam tilt out of band, and an input VSWR that varies widely with frequency.

In contrast, the branch feed configuration employs feed lines that are nominally of equal length. Therefore, the phase relationships of the radiating elements are maintained over a wide span of frequencies. This provides vertical patterns with stable beam tilt, a requirement for multichannel applications.

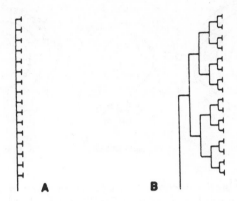

Figure 12-28 Antenna feed configurations: (a) series-feed, (b) branch-feed.

The basic building block of the multichannel antenna is the broadband panel radiator. The individual radiating elements within a panel are fed by a branch feeder system that provides the panel with a single input cable connection. These panels are then stacked vertically and arranged around a supporting spine or existing tower to produce the desired vertical and horizontal radiation patterns.

Bandwidth. The ability to combine multiple channels in a single transmission system depends upon the bandwidth capabilities of the antenna and waveguide or coax. The antenna must have the necessary bandwidth in both pattern and impedance (VSWR). It is possible to design an antenna system for low power applications using coaxial transmission line that provides whole-band capability. For high power UHF systems, waveguide bandwidth sets the limits of channel separation.

Antenna pattern performance is usually not a significant limiting factor. As frequency increases, the horizontal pattern circularity deteriorates, but this effect is generally acceptable. Also, the electrical aperture increases with frequency, which narrows the vertical pattern beamwidth. If a high gain antenna were used over a wide bandwidth, the increase in electrical aperture might make the vertical pattern beamwidth unacceptably narrow. This is, however, usually not a problem at UHF frequencies because of the channel limits set by the waveguide.

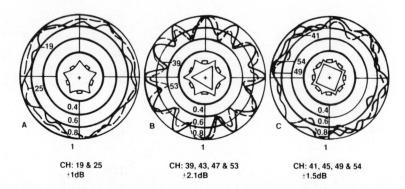

CH: 19 & 25
±1dB

CH: 39, 43, 47 & 53
±2.1dB

CH: 41, 45, 49 & 54
±1.5dB

Figure 12-29 Measured antenna patterns for three types of panel configurations at various operating frequencies: (a) 5 panels per bay; (b) 6 panels per bay; (c) 8 panels per bay. (*Source: Broadcast Engineering Magazine.*)

Horizontal Pattern. Because of the physical design of a broadband panel antenna, the cross-section is larger than a typical "narrowband" pole antenna. Therefore, as the operating frequencies approach the high end of the UHF band, the *circularity* (average circle to minimum or maximum ratio) of an omnidirectional broadband antenna generally deteriorates.

Improved circularity is possible by arranging additional panels around the supporting structure. Previous installations have used five, six, and eight panels per bay. These are illustrated in Figure 12-29 along with measured patterns at different operating channels. These approaches are often required for power handling considerations, especially when three or four transmitting channels are involved.

The flexibility of the panel antenna allows directional patterns of unlimited variety. Two of the more common applications are shown in Figure 12-30. The peanut and cardioid types are often constructed on square support spines (as indicated). A cardioid pattern may also be produced by side-mounting on a triangular tower. Different horizontal radiation patterns for each channel may also be provided, as indicated in Figure 12-31. This is accomplished by changing the power and/or phase to some of the panels in the antenna with frequency.

Most of these antenna configurations are also possible using a circularly polarized panel. If desired, the panel can be adjusted for elliptical polarization with the vertical elements receiving less than 50 percent of the power. Using a circularly-polarized panel will reduce the horizontally polarized ERP by half (assuming the same transmitter power).

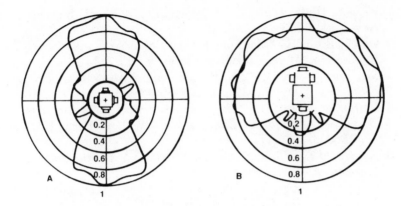

Figure 12-30 Common directional antenna patterns: (a) peanut, (b) cardioid. (*Source: Broadcast Engineering Magazine.*)

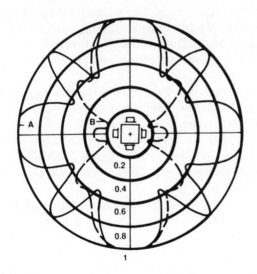

Figure 12-31 Use of a single antenna to produce two different radiation patterns: (a) omnidirectional; (b) peanut. (*Source: Broadcast Engineering Magazine.*)

Bibliography

1. Crutchfield, E.B., *National Association of Broadcasters Engineering Handbook*, Seventh Edition, Washington, D.C., 1985.

2. Benson, Blair, *Television Engineering Handbook*, McGraw-Hill, New York, 1986.

3. Chick, Elton B., "Monitoring Directional Antennas," *Broadcast Engineering Magazine*, Overland Park, KS, July 1985.

4. Mullaney, John H., "The Folded Unipole Antenna," *Broadcast Engineering*, Overland Park, KS, July 1986.

5. Mullaney, John H., "The Folded Unipole Antenna for AM Broadcast," *Broadcast Engineering*, Overland Park, KS, January 1960.

6. Raines, J.K., "Folded Unipole Studies." *Think Book Series*, Multronics, 1968–1969.

7. Howard, George P., "The Howard AM Sideband Response Method," *Radio World*, Falls Church, VA, August 1979.

8. Raines, J.K., "Unipol: A Fortran program for designing folded unipole antennas," Mullaney Engineering, Gaithersburg, MD, 1970.

9. Mullaney, John J., and George P. Howard, "SBNET: A Fortran program for analyzing sideband response and design of matching networks, Mullaney Engineering, Gaithersburg, MD, 1970.

10. Westberg, J.M., "Effect of 90° Stub on Medium Wave Antennas." *NAB Engineering Handbook*, Seventh Edition, Washington, D.C., 1985.

11. Mullaney, John H., P.E., "The Consulting Radio Engineer's Notebook," Mullaney Engineering, Gaithersburg, MD, 1985.

12. Mayberry and Stenberg, "UHF Multi-Channel Antenna Systems," *Broadcast Engineering*, Overland Park, KS, March 1989.

13. Anders, M.B., "A Case for the Use of Multi-Channel Broadband Antenna Systems," *NAB Engineering Conference Proceedings*, Dallas, TX, 1985.

14. Bingeman, Grant, "AM Tower Impedance Matching," *Broadcast Engineering*, Overland Park, KS, July 1985.

15. Bixby, Jeffrey, "AM DAs: Doing it Right," *Broadcast Engineering*, Overland Park, KS, February 1984.

16. Dienes, Geza, "Circularly and Elliptically Polarized UHF Television Transmitting Antenna Design," *Proceedings of the NAB Engineering Conference*, Las Vegas, NV, 1988.

17. Fink, D. and D. Christiansen, *Electronics Engineer's Handbook*, Third Edition, McGraw-Hill, New York, 1989.

18. Jordan, Edward C., *Reference Data for Engineers: Radio, Electronics, Computer and Communications*, Seventh Edition, Howard W. Sams Company, Indianapolis, IN, 1985.

19. Benson, B. and J. Whitaker, *Television and Audio Handbook for Technicians and Engineers*, McGraw-Hill, New York, 1989.

Index

About the Author

Jerry Whitaker is Associate Publisher of *Broadcast Engineering* and *Video Systems* magazines (Intertec Publishing Corp.). He is a Fellow of the Society of Broadcast Engineers, and an SBE-certified senior AM-FM engineer. He is also a member of the SMPTE, AES, ITVA and IEEE (Broadcast Society, Power Electronics Society, and Reliability and Maintainability Society). He has written and lectured extensively on the topic of RF transmission systems. Mr. Whitaker is a former radio chief engineer and television news producer. He is co-author of the McGraw-Hill *Television and Audio Handbook for Technicians and Engineers,* and a contributor to the McGraw-Hill *Audio Engineering Handbook.* Mr. Whitaker is also a contributor to the National Association of Broadcasters' *NAB Engineering Handbook, 7th Ed.*